Vanier Science

Physics in Canada
Book for Review

CAMBRIDGE MONOGRAPHS
ON MECHANICS AND APPLIED MATHEMATICS

GENERAL EDITORS

G. K. BATCHELOR, PH.D., F.R.S.
Professor of Applied Mathematics in the University of Cambridge

J. W. MILES, PH.D.
Professor of Applied Mechanics and Geophysics
University of California, La Jolla

SOUND TRANSMISSION THROUGH A FLUCTUATING OCEAN

SOUND TRANSMISSION THROUGH A FLUCTUATING OCEAN

STANLEY M. FLATTÉ (EDITOR)

Professor of Physics, University of California, Santa Cruz

ROGER DASHEN

Professor of Physics, Institute for Advanced Study, Princeton

WALTER H. MUNK

Professor of Geophysics, Scripps Institution of Oceanography, La Jolla

KENNETH M. WATSON

Professor of Physics, University of California, Berkeley

FREDRIK ZACHARIASEN

Professor of Physics, California Institute of Technology, Pasadena

CAMBRIDGE UNIVERSITY PRESS

CAMBRIDGE

LONDON · NEW YORK · MELBOURNE

Published by the Syndics of the Cambridge University Press
The Pitt Building, Trumpington Street, Cambridge CB2 1RP
Bentley House, 200 Euston Road, London NW1 2DB
32 East 57th Street, New York, NY 10022, USA
296 Beaconsfield Parade, Middle Park, Melbourne 3206, Australia

First published 1979

Printed in the United States of America

Typeset in Great Britain by J. W. Arrowsmith Ltd, Bristol BS3 2NT
Printed and bound by Vail-Ballou Press, Inc., Binghamton, New York

Library of Congress Cataloging in Publication Data
Main entry under title:

Sound transmission through a fluctuating ocean

(Cambridge monographs on mechanics and applied mathematics)
Bibliography: p. 277
Includes index
1. Underwater acoustics 2. Sound – Transmission
3. Oceanography I. Flatté, Stanley M.
QC242.2.S68   551.4′601   77-88676
ISBN   0 521 21940 X

TO
RENELDE
MARY
JUDITH
ELAINE
NANCY
AND
MIMI

I have long discovered that geologists never read each other's works, and that the only object in writing a book is a proof of earnestness, and that you do not form your opinions without undergoing labour of some kind.

Charles Darwin (1887)

CONTENTS

Sketch by Leonardo da Vinci   *Frontispiece*

Preface   xi

Acknowledgements   xiv

Introduction   xv

PART I. The ocean environment   1

1   Ocean structure   3
1.1   Scales   3
1.2   Water masses   10
1.3   Finestructure and microstructure   14
1.4   Circulation   19
1.5   The surface mixed layer   30
1.6   The canonical sound structure   31

2   Planetary waves and eddies   34
2.1   Planetary waves   35
2.2   Mesoscale   40
2.3   Geostrophic turbulence   42

3   Linear internal waves   44
3.1   Observed ocean fluctuations   44
3.2   Equations for internal-wave motion   46
3.3   Approximation to the wavefunctions $W(k, j, z)$ 53
3.4   The spectrum of internal waves   54
3.5   Equivalent spectra   57
3.6   The sound-speed correlation function   59

PART II. Introduction to sound transmission in the ocean   63

4   The ocean sound channel   65
4.1   Rays in the sound channel   65
4.2   Angle–depth diagrams   69

5   The wave equation   74
5.1   Fundamental approximations   74
5.2   The reduced wave equation and the parabolic approximation   76

5.3   Introduction to the path-integral formulation   78
5.4   Rays   82

PART III. Sound transmission through a fluctuating ocean   85

6   Transmission through a homogeneous, isotropic medium   87
6.1   Correlation functions and spectral functions   87
6.2   Parameters and regimes; Λ–Φ space   90
6.3   Geometrical optics   94
6.4   Other parameter regimes   99

7   The ocean medium   100
7.1   Fresnel zones and ray tubes   101
7.2   Definitions of the strength and diffraction parameters, Φ and Λ 106
7.3   The phase-structure function, D 108
7.4   Internal-wave dominance for Φ and Λ 110
7.5   Evaluation of the phase-structure function   117

8   Statistics of acoustic signals   120
8.1   Signal statistics and variables   120
8.2   Regimes in Λ–Φ space   126
8.3   One-point functions   130
8.4   Time separations   135
8.5   Spatial separations   139
8.6   Frequency separations   140
8.7   Pulse propagation   144

9   Multipath effects and n-point Gaussian statistics   150
9.1   Statistics of a wavefunction obeying n-point Gaussian statistics   150
9.2   Cartesian statistics   152
9.3   Intensity and phase statistics   154
9.4 n-point Gaussian statistics   158

PART IV. Theory of sound transmission   163

10   Supereikonal, or Rytov approximation   165
10.1 Isotropic ocean   165
10.2 Anisotropic ocean   170
10.3 Channeled ocean   174
10.4 Internal-wave dominance   175
10.5 Comparison with numerical experiments   185

11   Propagation through a single upper turning point   189
11.1 Setting up the problem   189
11.2 Regions in Λ–Φ space   193

11.3 Sound fluctuations in the presence of micromultipath 197
11.4 A better method of calculating sound fluctuations in the presence of micromultipath 200
11.5 Correlations in frequency 204

12 Path integrals and propagation in saturated regimes 207
12.1 The path integral 208
12.2 Signal statistics in the fully saturated regime 209
12.3 The Markov approximation 212
12.4 The partially saturated regime 217

13 The transport equation in sound scattering 220
13.1 The energy flux 221
13.2 The transport equation for acoustic intensity 222
13.3 The diffusion approximation 225
13.4 Scattering from internal waves 228
13.5 Scattering from the microstructure fluctuations 234

PART V. Experimental observations of acoustic fluctuations 237

14 Eleuthera–Bermuda 239
14.1 Treatment of data 239
14.2 Cartesian spectra 243
14.3 Phase and intensity statistics 248
14.4 Conclusions 248
14.5 The Williams and Battestin resolved experiment 250

15 Cobb seamount 252
15.1 Phase and intensity variances 252
15.2 Spectra 254

16 Azores 256
16.1 Environmental data 256
16.2 CW measurements 260
16.3 Pulse measurements 265

Epilog 269

Appendix A: Calculation of $K(\alpha)$ 273
Appendix B: Calculation of $Q(\alpha)$ 274
Appendix C: Calculation of γ 276

Bibliography 277
Glossary of terms 285
Units, dimensions and glossary of symbols 289
Index 295

PREFACE

A complex structure of motions, driven by winds, solar heat and tides, is continually at play in the ocean interior. The motions range from current and tidal patterns of planetary scale, through intermediate-scale processes like internal waves, to small-scale turbulence of sub-millimeter size. An appreciation of this complex and multi-connected system must form the basis of any understanding of the oceans, their physics, chemistry and biology, and of their interaction with the atmosphere above and the seafloor beneath. It must also form the basis of any purposeful intervention for the human good; whether it be for fishing, predicting weather and climate, or for communicating through the ocean.

The ocean is transparent to sound and opaque to electromagnetic radiation. Accordingly, there is a strong emphasis on acoustic methods in exploring the sea bottom, in locating fish, in communicating between ships and submarines, and in detecting vessels by active or passive sonar. 'If you let your ship stop, and dip the end of a long blowpipe in the water and hold the other end to your ear, then you can hear ships which are very far distant from you,' Leonardo da Vinci (1483) observed (see Frontispiece).

But even though conditions for acoustic transmission are remarkably favorable, the ocean structure sets the ultimate limit to what can be done. Ocean processes result in sound-speed fluctuations, typically $\delta C/C = 5 \times 10^{-4}$ in the upper layers, 3×10^{-6} at abyssal depth. These are small, yet they have a pronounced cumulative effect over long propagation paths. They prescribe the capacity of undersea communication systems, and impose a limit to the acoustic resolution of objects, similar to the resolution limit of ground-based telescopes commonly referred to as *atmospheric seeing*.

In the 1950s, physical oceanographers were occupied largely with drawing deterministic pictures of ocean circulation; they had very little information on the space-time variability of ocean structure. Only in the last five years have measurements of ocean variability been made that allow a semi-quantitative understanding of the oceanographic processes that are involved. At the planetary scales the ocean seems to be dominated by *geostrophic turbulence*, and at the intermediate scale by *internal waves*; in both instances horizontal scales vastly exceed vertical scales. For scales below one meter, little is known; concepts of isotropic turbulence, if valid anywhere, are confined to these very small scales. Time variations are not due to a frozen spatial structure being carried by a horizontal current, but are intrinsic to the processes themselves.

This information has been gathered painstakingly by the traditional tools of the physical oceanographer: bathythermographs, thermistors, salinometers, and current meters. These tools are beginning to pale somewhat before the task of measuring variations within the wide expanse of the sea. A long-range, integrating probe is needed, and measurements of sound propagation through the fluctuating ocean can provide such a probe.

Over the last century progress in the theory of wave propagation through fluctuating media has sprung from studies of electromagnetic wave propagation; particularly visible light in the atmosphere, and radio waves in interplanetary plasma. The techniques that have been developed include the Rytov method for treating weak diffraction, the transport equation for the energy flux, and the method of the propagation of moments. These methods have provided solutions to a wide variety of wave propagation problems, many of which are of practical importance in electromagnetic wave transmission. Unfortunately very few of these solutions apply to the ocean case, due to the special nature of the ocean sound-speed fluctuations. Some *ad hoc* attempts to adapt existing solutions so as to apply to the ocean case have been made, but the predictive power of such approaches is limited.

In this book, starting from general wave propagation equations, we attempt a systematic search for solutions that apply to the ocean medium. To characterize the ocean medium, we make rather extensive application of a model of internal waves, because it appears to portray observed ocean fluctuations in the important period range of minutes to a day, because it allows quantitative illustration of general principles, and because some existing acoustic experiments appear to be dominated by internal-wave effects. In special geographical regions, and for experiments with different scales, processes other than internal waves will dominate. Work on the effects of these other processes has barely begun.

In the process of deriving solutions applicable to the ocean case, we have developed a new and useful tool for treating wave propagation through random media; we call it the *micromultipath* technique. It has its roots in the principle of least action as expressed by Fermat and Hamilton, and it draws particularly on the path-integral technique of Feynman. It is to be hoped that this technique may prove useful in electromagnetic wave propagation also. However, the main purpose of the book is to provide a connection between the ocean structure as it is now evolving from oceanographic experiments, and the measured signal structure of sound transmitted through the oceans. We hope this will be of help to users of sound in coping with the frustrating fluctuations they continually encounter; we also hope this can be helpful in the inverse problem of using acoustic probes to monitor the everchanging ocean structure.

ACKNOWLEDGEMENTS

Much of the work presented here was carried out under the auspices of the Advanced Research Projects Agency (ARPA) of the United States Department of Defense. A large fraction of the work was completed during three summer studies of the Jason group in 1974, 1975, and 1976. The actual preparation of the book has been independent of ARPA, being partially supported by a Guggenheim Foundation Fellowship and a grant from the United States Office of Naval Research (ONR) to Professor Flatté. The European Center for Nuclear Research (CERN) extended Professor Flatté their hospitality during the period that much of the book was edited. Professor Munk gratefully acknowledges support by the ONR of his work on internal waves; this has provided the initial incentive for this work. We are grateful to Captain H. Cox, Professor F. Dyson, A. Ellinthorpe, and Dr F. Tappert for many fruitful discussions. We have benefited greatly from conversations with many workers in the acoustics and oceanography communities. We would particularly like to thank several experimental groups that generously allowed use of their data prior to publication. We thank Dr P. Wille of Kiel for drawing attention to the page in Leonardo da Vinci's notebook that provides the frontispiece. We acknowledge our debt to those who provided valuable comments on the text itself: H. Bezdek, H. Cox, A. Ellinthorpe, B. Lippmann, M. McKisic, P. Smith, B. Uscinski, and P. Worcester. We are grateful to the American Institute of Physics, Pergamon Press Ltd, John Wiley and Sons, Inc., Scientific American, Inc. and Prentice-Hall, Inc. for permission to reproduce figures in this book.

INTRODUCTION

Part I gives a description of the ocean environment through which sound is transmitted. The description is deliberately broad, from microstructure to planetary waves. Particular emphasis is placed on internal waves, for these form the basis of subsequent detailed calculations. But we wish to leave the reader with the clear realization that there is more to the oceans than internal waves. For example, it is well known that geographic variations are pronounced, and need to be taken seriously. Still, in a monograph such as this there is a need for a model ocean to provide an insight into the role played by various parameters, and to permit first-order comparison with experiment. The simplest model consistent with the warming in the upper kilometer at midlatitudes, and certain principles of water mass formation, is an ocean with a density gradient diminishing exponentially with depth. This same ocean model underlies both the shape of the ocean sound channel and the depth dependence of the internal-wave-induced sound-speed fluctuations.

Given an ocean with known structure, one method for obtaining sound-fluctuation predictions would be to create a computer-code simulation. We have done this, and the computer code has indeed provided a valuable tool for gaining insight into the character of sound-transmission fluctuations due to a particular ocean process, internal waves. However, any extensive computer code has its own approximations built in and is ultimately a rather cumbersome tool; accordingly, we have worked to develop analytical treatments for as many cases as possible.

Basic theoretical approaches to sound transmission in this book can be divided into two parts:

(1) The Rytov (or supereikonal) extension of geometrical optics valid in regions of *weak* scattering.

(2) A *micromultipath* theory capable of treating some aspects of *strong* scattering, and based on a formulation using path integrals.

These theories treat the fluctuations of a single deterministic sound-transmission path from a source to a receiver. However, it is well known that for transmission beyond a few tens of kilometers in deep water, sound travels along several well-separated paths simultaneously (see Chapter 4). The received signal on a single hydrophone in this *multipath* situation has additional fluctuations due to interference effects between paths. We have then a third area of theoretical treatment.

(3) *Macromultipath* effects – where the single path fluctuations can be added together to give predictions for multipath fluctuations.

The areas of application of these treatments can be understood more fully by referring to Fig. I.1: a range–frequency diagram. Transmission of sound in the sea to ranges within the area labelled 'absorption' is impractical due to the absorption of sound energy by seawater. Sound with frequencies below 10 Hz will be subject to effects of the surface and bottom acting as a waveguide. Between these two areas, treatments that are basically approximations or extensions to ray theory can apply.

The limit of validity of the supereikonal approximation is a function of the strength of sound-speed fluctuations. The line shown is a rough estimate of the limit of validity in the real ocean; the precise position of the line depends on factors other than range and frequency. The micromultipath area (also known as the *saturated* region) is basically that area where a single deterministic path can sporadically change into two or more paths due to the ocean fluctuations. Deterministic multipath effects are important for a single receiver beyond thirty kilometers, but a vertical array of hydrophones can separate the sound fluctuations from each path to a much larger range. (This is much easier in a computer than it is in the real ocean!) It is important to note that in much of the saturated region there are so many sporadic micropaths between source and receiver that *the essential sound characteristics are not altered by the additional imposition of deterministic multipath.*

Rays in the ocean are restricted to such small angles from the horizontal that the parabolic approximation to the wave equation is universally valid (Chapter 5). This approximation is the heart

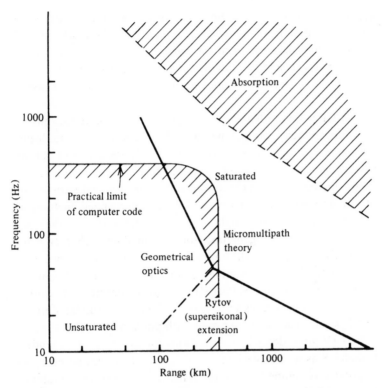

Fig. I.1. Limits and ranges of validity for various theoretical treatments discussed in this book. Seawater absorption of sound is greater than 20 dB in the shaded area.

of the computer code and is also used in most of the theoretical treatments.

In addition to these theoretical treatments we include a chapter on the transport equation, which treats scattering-angle distributions. This chapter provides a description of the connections between the micromultipath theory, the transport equation, and the method of the propagation of moments.

The description of all these theories of sound transmission is divided into three parts. Part II, an introduction to transmission theory, begins with a description of the most important feature in oceanic sound transmission: the sound channel (Chapter 4). In Chapter 5 a derivation of the wave equation and the parabolic approximation is given from first principles, and the particular method used in the computer code is described. This method leads naturally to a description of the path-integral formulation that is used in Parts III and IV. Chapter 5 concludes with a brief description of the geometrical-optics approximation and its region of validity.

Part III begins with a review of past work on sound transmission in homogeneous, isotropic media, and is important for establishing basic concepts and notation (Chapter 6). These concepts fall into two categories: (1) concepts required to characterize the sound–speed fluctuations; and (2) concepts of acoustic signal statistics and how to derive them from a knowledge of sound-speed fluctuations. Chapter 7 describes the fundamental differences between the ocean medium and a homogeneous, isotropic medium. Chapter 8 is a qualitative overview of the entire field of ocean sound-transmission fluctuations from the vantage points of the supereikonal and micromultipath theories. Part III ends with a description of multipath effects and includes a precise description of Gaussian statistics, which so often accompanies multipath phenomena (Chapter 9).

Part IV gives detailed justifications for the statements made in Chapter 8 as well as derivations of specific formulas based on the internal-wave model of Chapter 3. In addition, Part IV contains the transport-equation chapter relating the various wave propagation theories.

Part V compares the theories to field experiments at a variety of ranges and wavelengths. The reader who is experimentally oriented may find it most convenient to read Part III and then Part V, before plunging into the more detailed chapters in Part IV; the book has been organized to facilitate that order.

Previous books that have treated parts of ocean structure, particularly internal waves, in more detail than we have here are Phillips (1966) and Turner (1973). More recent progress is nicely summarized in an issue of the *Journal of Geophysical Research* devoted to internal waves. Previous books on sound transmission include Officer (1958), Tolstoy and Clay (1966) and Urick (1967 and 1975).

The quantitative connection between ocean structure and sound fluctuations has hardly been touched in the past. Phillips, Turner, and Officer do not consider the problem, while Urick and Tolstoy and Clay include very brief sections considering volume effects. Tolstoy and Clay view the ocean as a transmission channel, and describe its properties by certain correlation coefficients that are not readily identifiable with known physical processes. Urick, also in a brief section, considers quantitatively only isotropic inhomogeneities with a single scale size.

Other books of a more general nature, Brekhovskikh (1960 and 1975), Chernov (1960), and Tatarskii (1971) consider random media with a spectrum of scale sizes, but treat only isotropic cases without specializing to the ocean environment. Their results have been widely applied in ocean acoustics, but their neglect of anisotropy – so fundamental to the ocean environment – limits the usefulness of their results.

The present work is entirely devoted to the effects of fluctuating inhomogeneities in sound speed in the volume of the ocean. We admit that neglect of surface and bottom scattering is a fundamental difficulty for many practical problems; unfortunately, treatment of rough-interface scattering would require too much of an addition to an already thick volume. However, our knowledge that the ocean is nonisotropic (the scale sizes in the vertical being much smaller than in the horizontal) and that the strength of the inhomogeneities varies strongly with vertical position, has been taken into account in a basic way, as has the curved nature of unperturbed rays. These effects make the theoretical analysis more complicated, and the results are in many cases fundamentally different from those derived from homogeneous, isotropic turbulence theory.

PART I

THE OCEAN ENVIRONMENT

The following three chapters describe the average state of the ocean, its variability on a week-to-week and month-to-month basis (planetary waves and eddies), and its short-term variability (internal waves). The subsequent applications to ocean acoustics refer specifically only to the beginning of § 1.1, and §§ 1.6 and 3.6. This is because we place our emphasis (rather too heavily) on the role played by internal waves, and rely on a special model spectrum which illustrates our results by specific examples. Further, internal waves *are important*, as indicated in Part V by some striking agreements between measured and computed sound fluctuations.

But there is more to the ocean than internal waves! We have attempted in the first two chapters to give a description of the real ocean variability against which the application of a universal internal-wave spectrum can be judged, and to provide the background for future work involving the important effects of microstructure, ocean fronts, the surface mixed layer, and mesoscale variability. We make no pretense at completeness; this would involve all of oceanography. We totally omit a discussion of the surface and bottom boundaries and their role (sometimes crucial) to sound transmission. All biological factors are ignored.

The material in the first two chapters is descriptive, including dynamics only when it helps to systematize observed facts.

CHAPTER 1

OCEAN STRUCTURE

It used to be thought that the ocean, particularly the deep ocean, was uniform, but this was a case of mistaking lack of information for lack of features. In fact, the ocean waters are everywhere structured, with characteristic scales ranging from centimeters to the scale of ocean basins themselves. On the large scale the variability derives from an east–west asymmetry of the general circulation, and from distinct water masses produced in the Antarctic, Arctic, equatorial regions, and the marginal seas. But the production of mean-square gradients must be balanced on the average by an equivalent dissipation of mean-square gradients, and this inevitably leads to a cascade of scales into fine- and microstructure.

1.1. Scales

The oceans rotate and are stratified. The frequencies

$$\omega_i = 2\Omega \sin (\text{latitude}), \quad n = \left(-\frac{g}{\rho} \frac{d\rho}{dz} \right)^{\frac{1}{2}} \qquad (1.1.1)$$

are convenient measures of rotation and stratification. The inertial (or Coriolis) frequency ω_i is twice the vertical component of the Earth's angular velocity Ω. Accordingly, $\omega_i u$ is the Coriolis 'force' associated with a unit mass moving horizontally with speed u.

We will call n the buoyancy frequency (it is more often called the Brunt–Väisälä frequency). The fluid is statically stable when n is real, that is when the potential density increases with depth $-z$. A balloon filled with water at some depth and then displaced vertically will oscillate at a frequency n (under idealized conditions).

The inertial frequency increases from zero at the equator to $2\Omega = 2\pi/(12 \text{ hours}) = 1.46 \times 10^{-4} \text{ s}^{-1}$ at the poles, therefore the appropriate length scale is A, the radius of the Earth. The buoyancy frequency varies typically from 3 cph near the surface to 0.2 cph near the bottom; the fractional rate of decrease diminishes with depth, and can be roughly modeled by $n(z) = n_0\, e^{z/B}$, with B of order 1 km. Accordingly,

$$A = 6370 \text{ km}, \quad B \approx 1 \text{ km} \tag{1.1.2}$$

are the length scales associated with rotation and stratification respectively. Typical values for the frequencies are (at 30° latitude and 1 km depth)

$$\omega_i = \tfrac{1}{24} \text{ cph}, \quad n = 1 \text{ cph} \tag{1.1.3}$$

so that in this sense the Earth is a slowly spinning planet; ω_i/n (typically 0.04) is a measure of the aspect ratio (vertical to horizontal) of ocean inhomogeneities; there is no justification for the usual assumption of spherical symmetry.

The four scales ω_i, n, A, B are fundamental to the following discussion of ocean fluctuations. Planetary (or Rossby) waves have frequencies below ω_i. The ocean circulation is dominated by mesoscale eddies (associated with planetary waves) that have correlation distances and correlation times of order

$$2\pi Bn/\omega_i \approx 100 \text{ km}, \quad n^{-1}A/B = 1000 \text{ hours}$$

respectively (Chapter 2). Internal waves have frequencies lying between ω_i and $n(z)$. These are discussed in detail in Chapter 3.

Sound speed and stratification

It is convenient to introduce the *potential* gradient; defined as the measured gradient minus the adiabatic gradient, the latter arising from the adiabatic expansion or compression of a rising or sinking volume. Accordingly, the vertical gradient in density can be written as a sum of potential and adiabatic gradients,

$$\partial_z \rho = \partial_z \rho_P + \partial_z \rho_A,$$

with similar relations for temperature, T, and sound speed, C. Only the potential density gradient contributes to the stability of the water column. Similarly, only the potential gradient in sound speed, $\partial_z C_P$, contributes to sound fluctuations associated with internal waves and other forms of vertical motion. Aside from these fluctuations it is the true sound speed $C(z)$ that determines the properties of the sound channel.

Table 1.1. *Values of $10^6 a$ in deg^{-1}, at stated temperatures and depths. (From Munk, 1974.)*

Depth	Temperature (°C)					
(km)	0	5	10	15	20	25
0.0	51	107	158	204	245	283
1.0	76	127	173	216	254	290
2.0	100	146	189	227	263	296
3.0	122	164	203	238	271	302
4.0	143	181	217	249	279	308
5.0	161	197	229	258	286	314

Potential density is a function of potential temperature and salinity, $\rho_P(T_P, S)$, and so

$$n^2(z) = -g\rho^{-1} \partial_z \rho_P = g(a \, \partial_z T_P - b \, \partial_z S) \qquad (1.1.4)$$

where a and b are the coefficients of thermal expansion and saline contraction respectively. Similarly, the potential gradient in the fractional sound speed can be written

$$C^{-1} \partial_z C_P = \alpha \, \partial_z T_P + \beta \, \partial_z S. \qquad (1.1.5)$$

Typical numerical values are

$$\begin{aligned} \alpha &= 3.19 \times 10^{-3} \, (°C)^{-1}, \quad \beta = 0.96 \times 10^{-3} \, (\text{‰})^{-1} \\ a &\approx 0.13 \times 10^{-3} \, (°C)^{-1}, \quad b = 0.80 \times 10^{-3} \, (\text{‰})^{-1}. \end{aligned} \qquad (1.1.6)$$

The a-value is typical of conditions at 1 km depth, but varies considerably with temperature and pressure (Table 1.1).

The ratio

$$r = b \, \partial_z S / a \, \partial_z T_P \qquad (1.1.7)$$

gives the relative contributions of salt and (potential) temperature to the stability of the water column.

We define

$$G = \frac{\alpha}{a} \frac{s(r)}{g}, \quad \frac{\alpha}{a} = 24.5, \quad s(r) = \frac{1 + cr}{1 - r}, \quad c = \frac{a\beta}{\alpha b} = 0.049, \qquad (1.1.8)$$

using the numerical values (1.1.6). Then

$$C^{-1}\partial_z C_P = Gn^2(z). \qquad (1.1.9)$$

In any one location, G generally stays within relatively narrow limits, so that (1.1.9) expresses the potential sound-speed profile in terms of the stratification (which is perhaps the most fundamental aspect of ocean structure).

The total gradient in sound speed is the sum of the potential and adiabatic gradients,

$$\partial_z C = \partial_z C_P + \partial_z C_A,$$

where

$$
\begin{aligned}
C^{-1}\partial_z C_A &= \alpha\,\partial_z T_A + \gamma\,\partial_z P \\
&= (-0.03 - 1.11) \times 10^{-2}\,\text{km}^{-1} \\
&= -1.14 \times 10^{-2}\,\text{km}^{-1} \equiv -\gamma_A \qquad (1.1.10)
\end{aligned}
$$

is the fractional sound-speed gradient in an adiabatic, isohaline ocean. The total sound speed therefore increases with increasing T, S and P. Unlike C_P, C has a minimum value $C = C_1$ at some depth $z_1 = -h$ and increases by a few per cent towards top and bottom. The increase above the minimum is the result of increasing temperature towards the surface; the increase below the minimum is associated with increasing pressure in the nearly isothermal deep waters.

Vertical fluxes

What are the physical processes responsible for the stratification? Fig. 1.1 shows the interior distribution (ignoring the messy surface layers) of potential temperature and salinity in the eastern North Pacific. From the T–S diagram (Fig. 1.2) we obtain $r \approx -0.55$, so that a fraction $-r/(1-r) = 0.35$ of the density stratification is due to salinity.[†] At abyssal depths conditions are fairly uniform, characterized by 1.1 °C potential temperature and 34.69 ‰ salinity. This is Antarctic Circumpolar Water spreading northward into the deep-basin of the world's oceans. The formation of new bottom water is accompanied by upwelling throughout the water column. We can

---

[†] At 40° S in the Pacific, $r \approx -0.44$. In the North Atlantic the salinity partially destabilizes; a typical value for Bermuda is $r = +0.8$.

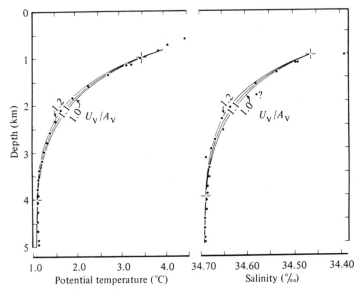

Fig. 1.1. Potential temperature and salinity as functions of depth at station CALCOFI 60–190, 33° 17′ N, 132° 42.5′ W (salinity at depth 1859 m was questioned in the original observations). Curves are labeled in reciprocal scale depth, units km^{-1}.

postulate (rather naively) that the observed vertical distribution represents a balance between vertical convection and diffusion:

$$[A_V \, d^2/dz^2 - U_V \, d/dz] \, T, S = 0, \qquad (1.1.11)$$

with temperature and salinity fixed at top and bottom. For constant vertical eddy diffusivity, A_V, and upwelling, U_V, this leads to an exponential distribution with depth scale $A_V/U_V \approx 1$ km (Fig. 1.1). In the case of nonconservative quantities (such as ^{14}C) we need to include the decay constant D (1.24×10^{-4} years^{-1}) as an additional term within the brackets of (1.1.11), leading to a time scale $A_V/U_V^2 \approx 200$ years (Munk, 1966). Hence, with A_V/U_V and A_V/U_V^2 given,

$$A_V \approx 1 \text{ cm}^2 \text{ s}^{-1}, \quad U_V \approx 1 \text{ cm day}^{-1} = 5 \text{ km}/(1000 \text{ years}).$$
$$(1.1.12)$$

This model predicts a complete cycling of ocean water every one thousand years.

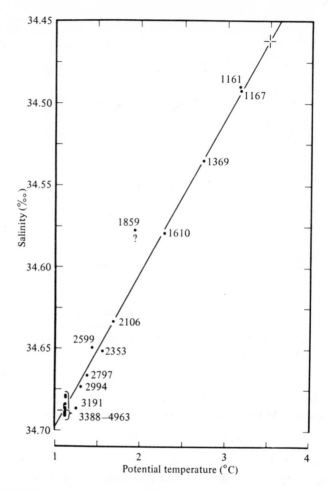

Fig. 1.2. The potential temperature versus salinity relation for station CALCOFI 60–190, with depth indicated in meters.

There are many things wrong with this picture of a vertical diffusive–convective model driven by the formation of Antarctic Bottom Water. Other water masses intrude at intermediary depths and modify the assumption of a constant U_V. The parameterization of vertical diffusion by a constant A_V is certainly inadequate, though the processes are not yet understood. (Leading theories

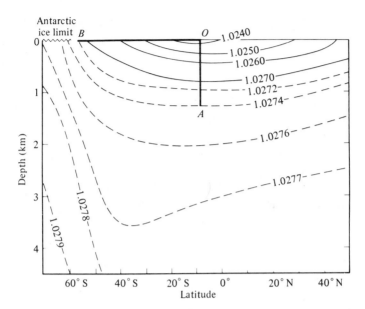

Fig. 1.3. Schematic representation of potential density in a north–south section of the Pacific.

include double diffusion discussed in Turner, 1973; and internal-wave breaking.) Finally, we have ignored all lateral processes.

Lateral fluxes

Referring to the schematic presentation of Fig. 1.3, we find that the T–S relation along the vertical section OA is very similar to the section OB along the sea surface. (This mapping of latitude into depth was noted by Iselin, 1939, in the Sargasso Sea.) The interpretation is that the water attains T–S characteristics at the air–sea boundary, and these are then readily communicated into the interior along surfaces of constant potential density. The profile of $\rho = 1.0274$ emerges just north of the antarctic ice limit, and we may expect the water above this level (about 1 km) to be strongly influenced by surface heating and cooling, precipitation and evaporation. This is discussed in the following section. There is no good agreement as to the relative importance of vertical and horizontal mixing processes.

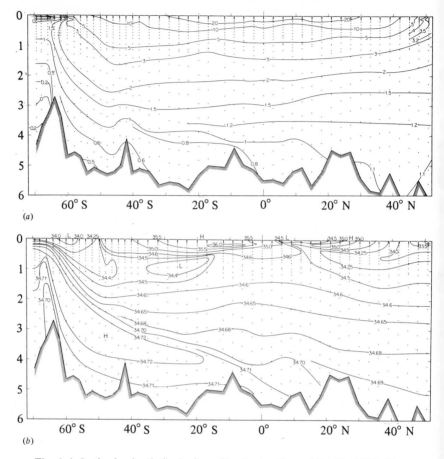

Fig. 1.4. Latitude–depth (in km) profiles in the Central Pacific (170° W), from Antarctica to the Aleutians. (a) Potential temperature contour in °C; (b) salinity contour in ‰; (c) sound-speed contour (from 1450 to

1.2. Water masses

Latitudinal variation of ocean structure is illustrated by an antarctic to arctic section through the Central Pacific (Fig. 1.4[†]). Observations north of 34° N were taken during the *Great Bear Expedition*, moving southward, April to June 1970 (Horibe, 1971); the remain-

---

[†] Joseph Reid has selected the data for the construction of these sections. We are grateful for his help and advice.

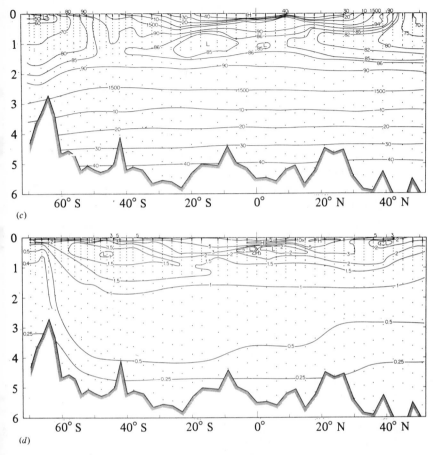

(c)

(d)

1540 m s^{-1}, the first two digits are omitted); and (d) buoyancy frequency contour in cph. Vertical exaggeration is about 1000:1; dots give measurement grid.

ing observations were taken during the *Southern Cross Cruise*, moving southward, November 1968 to March 1969 (Horibe, 1970).

Bottom water south of the Equator is characterized by about 0.6 °C, 34.71 ‰. This water is a derivative of the Antarctic Bottom Water which is formed mostly within the Weddell Sea by freezing processes (not yet completely understood) that generate a residual of salty, cold and therefore dense water (Gill, 1973). By the time it has moved (within the Antarctic Circumpolar Current) into the

South Pacific, it has undergone some mixing with the overlying waters and is warmer, more saline, and less dense.

Above the Antarctic Bottom Water is a saltier and warmer water mass, the Antarctic Circumpolar Water, reaching a maximum of 34.74 ‰. There is no process by which this water could have formed entirely in the Weddell Sea. Reid and Lynn (1971) have traced the origin of the Antarctic Circumpolar Water to the very saline and cold waters in the Norwegian–Greenland Seas. It then moves southward along the surfaces of equal density to form the North Atlantic Deep Water. These surfaces rise to shallow depths at high southern latitudes, and there the waters move eastward around the Antarctic Continent until they mix with waters from the Weddell Sea. The Antarctic Circumpolar Water so formed penetrates the Pacific Ocean, and is recognizable by a salinity maximum all the way to the Equator. Even at 33 °N (Figs. 1.1, 1.2) the deepest water has a T–S relation appropriate to Antarctic Circumpolar Water.

Higher up in the southern oceans, at about 1 km depth, is a salinity minimum which can be traced well north of the Equator. This characterizes the Antarctic Intermediate Water, formed by 2.2 °C and 33.8 ‰ surface water sinking at the antarctic convergence[†] (55° S) and mixing with surrounding water. A corresponding salinity minimum north of the Equator is associated with the Subarctic Intermediate Water.

To summarize, we have in the Pacific

Water type	Mean depth at 45° S (km)	Potential temperature (°C)	Salinity (‰)
Antarctic Intermediate Water	1	5	34.4 (min)
Antarctic Circumpolar Water	3	1.5	34.74 (max)
Antarctic Bottom Water	5	0.6	34.71

[†] Fronts and convergences are discussed in §1.4.

The reader will recognize these water types for what they are: convenient abstractions in a complex convective–diffusive equilibrium.

Turning now to surface processes, we find zones of high surface salinity at latitudes 20° to 30° on both sides of the Equator, corresponding to regions of high evaporation and low rainfall. Further north the situation is reversed: a salinity minimum corresponds to maximum rainfall and minimum evaporation.

The field of potential temperature is simpler. At great depth the temperature is typically 1 °C. The deep southern ocean is colder, reaching slight negative values at extreme southern latitudes. A temperature minimum (<-1 °C) is found near the surface at 65 °S. (The freezing point of seawater is well below 0 °C, typically about -1.8 °C.) At shallow depths a wedge of warm water (>20 °C) extends from 30° S to 30° N. Thus, the low latitudes are warm and fresh, the middle latitudes are moderate and salty, and the high latitudes are cold and fresh. These properties are communicated to the interior along surfaces of constant potential density.

Sound speed increases with temperature, salinity and pressure. In low- and midlatitudes a sound-speed minimum of about 1485 m s^{-1} at about 1 km depth (the sound-channel axis) reflects a balance between the effects of increasing temperature above, and increasing pressure below. The sound speed reaches about 1540 m s^{-1} at top and bottom. That these values should be so nearly alike is surely an accident, but this accident has important consequences on sound propagation.

The latitudinal variation is largely controlled by the corresponding temperature field, and less so by salinity. Thus, the sound channel terminates south of 55° S because of the rapid temperature and salinity drop across the antarctic convergence. A similar role is played by the subarctic front at about 45° N. As a consequence of higher temperature and salinity to the south, at southern midlatitudes the sound-channel axis is somewhat deeper and the minimum speed higher than at northern midlatitudes. Abyssal sound speeds and temperatures are higher to the north. The lowest speed (1446 m s^{-1}) is found in the shallow antarctic temperature minimum.

Table 1.2. *Comparison of properties of the upper one kilometer of the Atlantic and Pacific Waters.* (*Reid, J.*, 1961.)

	Temperature (°C)	Salinity (‰)	Potential density (g cm^{-3})	Sound speed (m s^{-1})
Atlantic				
North	11.78	35.42	1.02698	1505.5
South	7.73	34.69	1.02710	1489.7
Mean	9.58	35.02	1.02707	1497.1
Pacific				
North	9.58	34.42	1.02660	1496.3
South	9.51	34.67	1.02681	1496.4
Mean	9.54	34.55	1.02670	1496.3

The field of buoyancy depends on the vertical *gradients* of temperature and salinity, and is therefore more complex than the preceding plots. An equatorial saddle point is associated with the Equatorial Undercurrent, a jet-like current flowing westward along the Equator opposite to the direction of surface motion. At great depths the computed values are uncertain, being based on very small vertical differences. We can recognize the role of the antarctic convergence and subarctic front, as well as a shallow effect of the subtropical front at 30° N. For a typical midlatitude section the buoyancy frequency is 5 cph at 0.3 km and 0.25 cph at 4.5 km, decreasing by a factor e with each 1.4 km of depth.

There are, of course, substantial distinctions between ocean basins. For example, the North Atlantic is warmer, saltier and denser than the North Pacific (Table 1.2). The north–south asymmetry of the Atlantic is pronounced, not at all like the Pacific section in Fig. 1.4. An explanation for these differences is not readily available, but they are clearly related to the configuration of the basins.

1.3. Finestructure and microstructure

Contours in Fig. 1.4 are based on classical hydrographic casts employing reversing (Nansen) bottles typically at hundred-meter intervals in the upper oceans, and at half-kilometer intervals at abyssal depths. Only the gross features can be resolved in this way. The development of continuously sounding 'modern' instruments (BT, STD, CTD) has demonstrated a temperature and salinity structure down to meter scale. This was subsequently extended to centimeter scale by free-fall apparatus sinking slowly ($\sim 0.1 \, \text{m s}^{-1}$) and employing rapid-response (~ 0.01 s) transducers (Gregg and Cox, 1972; Gregg et al., 1973; Gregg, 1975). As a result of these instruments the temperature structure has now been resolved. The salinity structure goes to even smaller scales and has not been resolved.

The evolving terminology

> gross structure: larger than 100 m
> finestructure: 1 m to 100 m
> microstructure: less than 1 m

is based largely on what could be resolved in a given epoch by existing apparatus. But there is some evidence for a physical transition between the fine- and microscales: temperature spectra show a kink at 1 cpm, with wavenumbers above 1 cpm falling off less rapidly, if at all. Further, the Richardson scale $(\varepsilon/n^3)^{\frac{1}{2}}$ is of order 1 m for typical values of the dissipation per unit mass ($\varepsilon = 10^{-4} \, \text{cm}^2 \, \text{s}^{-3}$) and buoyancy frequency ($n = 10^{-3} \, \text{s}^{-1}$); for vertical scales large compared to the Richardson scale, buoyancy plays a dominant role.

Isotropic turbulence is relegated in the oceans to scales smaller than 1 m. It has been suggested that temperature and salinity microstructure is sometimes the residue from a turbulent event (fossil turbulence) after the velocity microstructure has dissipated. The microstructure is patchy in space and time; early descriptions, based on the pioneering observations of J. Woods (1968) in the Mediterranean, referred to 'sheets and layers' in the finestructure (see also Grant et al., 1968).

The generation of fine- and microstructure is not well understood; whatever the processes, the existing ocean structure reflects a balance between the formation and dissipation of mean-square gradients. The microstructure, in which most of the mean-square gradients (and hence dissipation) reside, plays a vital function.

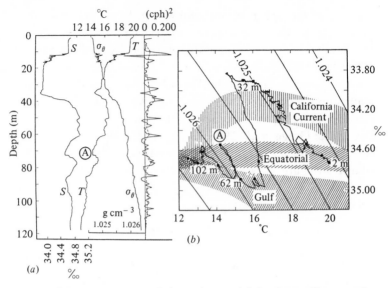

Fig. 1.5. (*a*) Temperature, salinity and potential density profiles at station MR7. The 'raw' n^2 profile is averaged over 0.8 m intervals (solid), and a 'smooth' n^2 profile is averaged over 3 m intervals (dashed). (*b*) The associated T–S diagram, with shaded areas indicating the three principal water masses off Cabo San Lucas. Lines of equal potential density are indicated. (From Gregg, 1975.)

Fig. 1.5 illustrates a complex *intrusive* situation 60 km southwest of Cabo San Lucas, the southern tip of Baja California. Three water masses intermingle: the saline outflow from the Gulf of California, the relatively fresh waters being brought in from the northwest by the California Current, and equatorial water of intermediate salinity from the eastern tropical Pacific. Microstructure record MR7 illustrates the intermingling of three water masses, each jostling for a level appropriate to its density. Temperature inversions (dT/dz negative) are balanced by positive salinity gradients, so that the density increases with depth ($-z$), and n^2 is generally positive. The inversions have typical vertical scales of 5 m, with a step structure (e.g., feature A in Fig. 1.5) attributed to the *diffusive* regime of double-diffusive[†] convection. The undersides of

[†] Double-diffusive instabilities are associated with the differences in the rates of diffusion of heat and salt (see Chapter 8 in Turner, 1973).

Table 1.3. *Rms and mean temperature gradients off Cabo San Lucas and in the central North Pacific. (From Gregg, 1975.) Units are* °C m^{-1}. *The symbol* Δ *is the fractional variance in temperature gradient defined in* (1.3.1).

	Station		
	MR7	MR6	MSR4
$\langle(\partial_z T)^2\rangle^{\frac{1}{2}}$	5.100	0.097	0.038
$\langle\partial_z T\rangle$	0.067	0.028	0.026
Δ	5740	12	2

the temperature inversions are often characterized by strong salinity inversions, and by prominent microstructure presumably attributed to the *fingering* regime of double-diffusive convection. Ultimately, the double-diffusive processes must lead to the destruction of the intrusive features.

At a distance of only a few kilometers from MR7, the record MR6 was characteristic of Equatorial Water throughout its depth interval, and the structure is much smoother. This is conveniently portrayed by the fractional variance in temperature gradient,

$$\Delta = (\partial_z T - \langle\partial_z T\rangle)^2/\langle\partial_z T\rangle^2, \qquad (1.3.1)$$

which varies by a factor of 500 between the stations (Table 1.3). For comparison we have included MSR4 for the mid-gyre of the central North Pacific (Gregg *et al.*, 1973), where the fine- and microstructures are almost absent. These stations characterize strongly intrusive, weakly intrusive and non-intrusive situations. The study of these intrusive features is just beginning, but it is clear that broad generalizations will be hard to come by.

The contribution to the mean-square gradients by different vertical wavenumbers is portrayed in Table 1.4. Spectral levels diminish up to 1 cpm, and then level off (particularly for MR7). This

Table 1.4. *Vertical wavenumber spectra in the vertical gradients of temperature, salinity and potential density at three stations.*

	Station								
	MR7			MR6			MSR4		
cpm	0.1	1	10	0.1	1	10	0.1	1	10
$\left(\dfrac{°C}{m}\right)^2 \Big/ \text{cpm}$	2×10^{-1}	1×10^{-2}	1×10^{-2}	2×10^{-3}	2×10^{-4}	7×10^{-5}	7×10^{-4}	1×10^{-4}	3×10^{-5}
$\left(\dfrac{‰}{m}\right)^2 \Big/ \text{cpm}$	2×10^{-2}	1×10^{-3}	8×10^{-4}	5×10^{-5}	1×10^{-5}	?	4×10^{-6}	8×10^{-6}	2×10^{-5}
$\left(\dfrac{\text{g cm}^{-3}}{m}\right)^2 \Big/ \text{cpm}$	2×10^{-9}	7×10^{-10}	7×10^{-10}	2×10^{-10}	2×10^{-11}	?	2×10^{-11}	1×10^{-11}	9×10^{-12}?

may be an indication of a physical distinction between finestructure and microstructure processes.

The relative contribution to the density spectra from temperature and salinity can be estimated. Let

$$\rho^{-1}\, \partial_z\rho = -a\, \partial_z T + b\, \partial_z S,$$

with a and b designating the coefficients of thermal expansion and saline contraction (values are given in (1.1.6)). For MR7 at 0.1 cpm, the measured density spectrum is much smaller than that inferred from either temperature alone, or salinity alone (Table 1.5). The implication is that temperature and salinity effects on density nearly cancel as they do for the gross structure (Fig. 1.4). This is consistent with a correlation between warm and salty anomalies, and between cold and fresh water anomalies, observed by Gregg (1975). At high wavenumbers for MR7, and at all wavenumbers for MR6 and MSR4, the density spectrum is of the same order or larger than that inferred from the temperature and salinity fluctuations. The implication is that intrusive features dominate MR7 at the 10 m scale, but not at smaller scale, and that intrusive features do not dominate at any scale for the other stations. We surmise that the non-intrusive structure with vertical scales of 1 to 200 m is related to internal-wave activity, but this issue has not been settled.

Table 1.6 gives the finestructure and microstructure in sound velocity, as inferred from Tables 1.3 and 1.4 according to

$$C^{-1}\, \partial_z C_{\mathrm{P}} = \alpha\, \partial_z T + \beta\, \partial_z S.$$

Values of α and β are given in (1.1.6). The temperature effect is clearly dominant.

The horizontal scales exceed the vertical scales by an aspect ratio $1000:1$ to $100:1$ for the finestructure, decreasing perhaps to order unity for 10 cm microstructure. These ratios are not well known.

1.4. Circulation

Steady currents play a strong role in the distribution of water masses. The overriding factor is that the flow is directed *along* isobars (rather than 'downhill' across isobars) in accordance with

Table 1.5. *Measured spectra in potential density, and those associated with temperature and salinity structure. Units are* (g cm^{-3} m^{-1})^2 cpm^{-1} .

	Station								
	MR7			MR6		MSR4		
cpm	0.1	1	10	0.1	1	0.1	1	10
Spectrum of $\rho^{-1}\partial_z\rho$	2×10^{-9}	7×10^{-10}	7×10^{-10}	20×10^{-11}	20×10^{-12}	2×10^{-11}	10×10^{-12}	9×10^{-12}?
Spectrum of $a\,\partial_z T$	8×10^{-9}	4×10^{-10}	4×10^{-10}	4×10^{-11}	8×10^{-12}	3×10^{-11}	4×10^{-12}	1×10^{-12}
Spectrum of $b\,\partial_z T$	13×10^{-9}	6×10^{-10}	5×10^{-10}	3×10^{-11}	6×10^{-12}	0.3×10^{-11}	5×10^{-12}	13×10^{-12}

Table 1.6. Rms values, mean values and spectra in the gradient of relative sound velocity, $C^{-1} \partial_z C$, as inferred from Tables 1.3 and 1.4. Gradient units are in m^{-1}, spectra in $(m^{-1})^2$ cpm^{-1}.

	Station								
	MR7			MR6			MSR4		
$\alpha \langle (\partial_z T)^2 \rangle^{\frac{1}{2}}$	1.6×10^{-2}			3.1×10^{-4}			1.2×10^{-4}		
$\alpha \langle \partial_z T \rangle$	2.1×10^{-4}			8.9×10^{-5}			8.3×10^{-5}		
cpm	0.1	1	10	0.1	1	10	0.1	1	10
Spectrum of $\alpha \partial_z T$	2×10^{-6}	1×10^{-7}	1×10^{-7}	2×10^{-8}	2×10^{-9}	7×10^{-10}	7×10^{-9}	1×10^{-9}	3×10^{-10}
Spectrum of $\beta \partial_z S$	2×10^{-8}	1×10^{-9}	7×10^{-10}	5×10^{-11}	1×10^{-11}	?	4×10^{-12}	7×10^{-12}	2×10^{-11}

the *geostrophic relation*

$$(2\Omega \sin \phi)v = (A \cos \phi)^{-1}(\partial p/\partial \lambda), \quad (2\Omega \sin \phi)u = -A^{-1}(\partial p/\partial \phi)$$
$$(1.4.1)$$

where λ, ϕ are east longitude and north latitude, u, v are eastward and northward components of flow, and $dp/d\lambda$, $\partial p/\partial \phi$ the corresponding components in horizontal pressure gradient. A, Ω are the Earth's radius and angular velocity. The geostrophic relation holds at low frequencies $\omega \ll \omega_i = 2\Omega \sin \phi$ for which the acceleration terms du/dt, dv/dt are relatively small. (For internal waves $\omega \gtrsim \omega_i$.) In the case of impulsive generation, geostrophic flow is essentially established after a few inertial oscillations.

Equation (1.4.1) is not to be interpreted in a cause-and-effect way; it applies equally well to both *thermohaline* circulation and *wind-driven* circulation. The thermohaline circulation is due to the formation of water masses of varying densities. The wind-driven circulation is the direct effect of surface forcing. In either case we end up with a geostrophic balance between the Coriolis and pressure forces according to (1.4.1).

For an exterior pressure field imposed upon the surface, the geostrophic relation gives a depth-independent current (*barotropic* flow). Of more importance is the pressure field associated with the interior distribution of temperature and salinity, giving rise to depth-dependent *baroclinic* currents. The pressure field is related to the density field through the hydrostatic approximation

$$\partial_z P = -\rho g, \tag{1.4.2}$$

and the density field to temperature, salinity and pressure by the equation of state

$$\rho = \rho(T, S, P), \tag{1.4.3}$$

so that there is an intimate connection between the circulation and the distribution of temperature and salinity in the oceans.

The equations (1.4.1), (1.4.2) and (1.4.3) together with the flux equations for heat, salt and mass form a system of seven equations in the unknowns u, v, w, p, ρ, T, S. The difficulty in deriving the distribution of ocean variables from fundamental principles is of two kinds: (i) the convective transports of heat and salt impose a nonlinear coupling that requires the combined treatment of water mass formation and current generation even in the lowest order treatment; and (ii) small-scale processes (waves, convection, turbulence, . . .) have to be parameterized to provide for their vital role in the gross distribution. This parameterization is not easily accomplished, particularly when some of the vital processes have not even been identified. The usual way out is to resort to diffusive processes with vertical and horizontal eddy coefficients A_V, A_H which are then adjusted to fit the data, unfortunately they turn out to be enormously variable. In spite of all these difficulties, there has been progress in ocean modeling. We refer in particular to the analytical modeling of Veronis (1969) and of Rattray and Welander (1975) and to the numerical modeling of Bryan, Manabe and Pacanowski (1975). In the analytical papers the authors find an exponential ocean $n = n_0 \exp(z/B)$ to be a solution (though not the only one), and they end up with scales for vertical eddy diffusivity A_V and vertical convective velocity U_V that are not totally out of line with those arising from the greatly oversimplified vertical flux model (1.1.12).

In the following discussion of thermohaline and wind-driven circulations their strong interdependence is ignored.

Abyssal circulation

In his study on the *Meteor* observations Wüst (1955) found that the bottom water from the Antarctic made its way northward as a narrow current hugging the western boundary of the South and North Atlantic basins. At somewhat shallower depth, North Atlantic Deep Water flowed southward, again hugging the western boundary. This picture was rather different from the previous interpretation in which the deep and bottom waters moved as continuous sheets across the full width of the ocean basin.

Geostrophy imposes certain restraints upon circulation with implications that are not easily visualized. For fixed east–west pressure gradients, the associated north–south flow must vary with latitude because of the variation in the Coriolis force and the convergence of longitudes, requiring a very special distribution of sources and sinks to accommodate conservation of mass. East–west flow is permissible without divergence, except when blocked by boundaries.

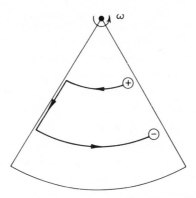

Fig. 1.6. Diagram and circulation induced in a rotating sector by source $\oplus$ and sink $\ominus$ positioned as shown. (From Stommel *et al.*, 1958.)

For a simple model with point source and sink near an eastern boundary, a geostrophic analysis by Stommel *et al.* (1958) led to the flow indicated in Fig. 1.6, and they subsequently confirmed this result by experiment in a rotating tank. In the absence of any distributed sink, the only permissible flow is westward to the boundary, then along the boundary in a narrow intense current, and finally eastward to the sink. The flow pattern is circuitous compared to what occurs in the absence of rotation. In the same spirit, Stommel (1958) modeled the abyssal circulation of the world's oceans (Fig. 1.7), taking two sources of $20 \times 10^6 \, \mathrm{m^3 \, s^{-1}}$ each, one in the North Atlantic and one in the Weddell Sea, with a distributed sink of $40 \times 10^6 \, \mathrm{m^3 \, s^{-1}}$ at shallow depth. This corresponds to an upward velocity of order 1 cm day^{-1} (1.1.12). The principal features are in accord with Wüst's bottom temperature maps and with subsequent observations. In particular, the salinity maximum extending northward into the deep South Pacific (Fig. 1.4) can be traced from the surface water of the Norwegian–Greenland Sea, southward along the coast of America, and then eastward halfway around the world in the Antarctic, into the Pacific (Reid and Lynn, 1971).

Wind-driven circulation

The surface currents of the North Pacific are shown in Fig. 1.8. Velocities diminish with depth with a typical scale of 1 km, and so it

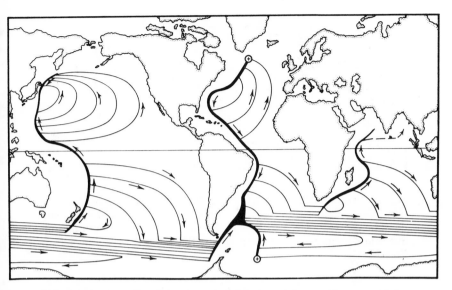

Fig. 1.7. A schematic of the abyssal circulation of the world oceans. If the two sources $\oplus$ in the North and South Atlantic are taken as $20 \times 10^6 \, \mathrm{m}^3 \, \mathrm{s}^{-1}$ each, then transport between streamlines is of the order of $5 \times 10 \, \mathrm{m}^3 \, \mathrm{s}^{-1}$. (From Stommel, 1958.)

is convenient to represent the circulation by vertically integrated transport. The most characteristic feature is the subtropical gyre between the westward equatorial current to the south, and the eastward drift to the north. The gyre has west–east asymmetry with a concentration of flow in the Western Boundary Current (Kuroshio). The situation is similar in the North Atlantic, with the Gulf Stream as Western Boundary Current.

A generalized circulation is shown in Fig. 1.9. The boundaries of the gyres fall along the latitudes of the major wind systems. Regardless of the gyre's sense of rotation, and regardless of whether the schematic is applied to the northern or southern hemisphere, the current intensification is always along the *western* boundary.

In a general sense the circulation pattern is understood. Using the stream function ψ of vertically integrated velocity u as an independent variable,

$$\int_{-h}^{0} \mathbf{u} \, dz = \hat{\mathbf{z}} \times \nabla \psi, \qquad (1.4.4)$$

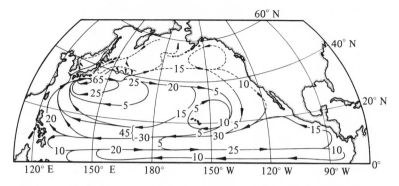

Fig. 1.8. Transport chart of the North Pacific. Numbers give indicated transport volumes above 1500 m in units of 10^6 m s^{-1}. (There is now some question concerning the existence of the gyre northeast of Hawaii.) (From Sverdrup, Johnson and Fleming, 1970. Reproduced by permission of Prentice-Hall, Inc.)

where $\hat{z}$ is a vertical unit vector, and h the water depth. The equations of motion can be represented as a balance of three torques:

$$\beta\, \partial_x \psi + A_V \nabla^2 \psi = \operatorname{curl} \tau. \qquad (1.4.5)$$

The first term is the planetary vorticity associated with northward transport $M_y = \partial_x \psi$ and the northward derivative $\beta = \mathrm{d}f/\mathrm{d}y$ of the Coriolis force $f = 2\Omega \sin \phi$ where x is the eastward and y the northward coordinate. The second term is the vorticity associated with a frictional force proportional to u ($A_H \nabla^4 \psi$ for lateral friction). The third term is the vertical component of wind-stress torque. Nearly everywhere the balance is between planetary vorticity and wind-torque, giving rise to the *Sverdrup transport*, $M_y = \beta^{-1} \operatorname{curl} \tau$, except in a Western Boundary Current of width $(A_H/\beta)^{\frac{1}{3}}$, where the balance is essentially between planetary and frictional vorticity.[†] The net transport of the Western Boundary Current can be computed by integrating the wind-stress torque across the ocean basin,

$$M = \beta^{-1} \int \operatorname{curl} \tau \, \mathrm{d}x, \qquad (1.4.6)$$

[†] This is the situation also for the Western Boundary Current in the thermohaline circulations.

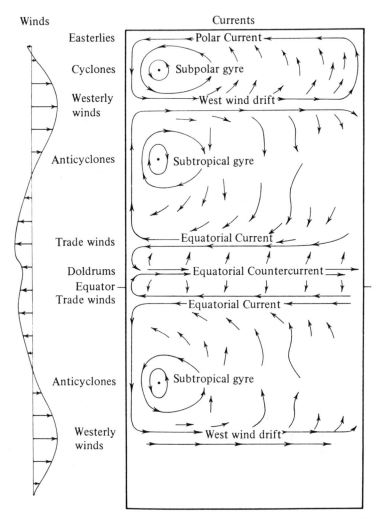

Fig. 1.9. Schematic representation of circulation in a rectangular ocean. (From *The circulation of the oceans*, W. H. Munk. Copyright © 1955 by Scientific American, Inc. All rights reserved.)

and this leads to the observed order of magnitude. There is some difficulty in sorting the thermohaline from the wind-driven transport in the observations, and the linear theory is demonstrably inadequate to deal with the jet-like features of the Western Boundary Currents, particularly where they decelerate and turn seaward.

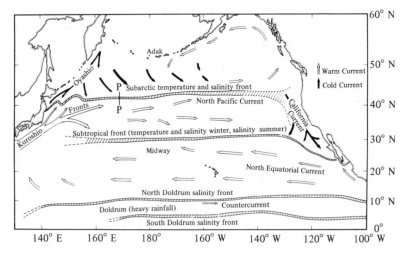

Fig. 1.10. Schematic map of the main North Pacific fronts. A vertical section across the subarctic front at *P–P* is shown in Fig. 1.11. (From Roden, 1975.)

The circulation around the Antarctic Continent differs from that in the bounded ocean basins. The flow is essentially eastward around the entire Earth, driven directly by the winds and attaining a transport in excess of $100 \times 10^6 \, \text{m}^3 \, \text{s}^{-1}$. The restriction of Drake Passage has, however, important dynamic considerations (Kamenkovich, 1962; Gill and Bryan, 1971).

The Equatorial Undercurrent (or Cromwell Current) flows eastward along the Equator in all oceans. It attains maximum velocities of $1 \, \text{m} \, \text{s}^{-1}$ at 100 m depth. It is approximately 200 km across, from 1° S to 1° N. The total transport is a surprising $40 \times 10^6 \, \text{m}^3 \, \text{s}^{-1}$. The dynamics of this equatorial singularity have been extensively investigated (for example, Gill, 1971), but the problem has not been settled; in all events, the Equatorial Undercurrent plays a decisive role in the equatorial distribution of ocean parameters.

Fronts

Returning to Fig. 1.9, we note that conditions within a gyre tend to be relatively uniform. The subtropical gyre, for example, encloses a relatively warm and saline body of water, poor in phosphates,

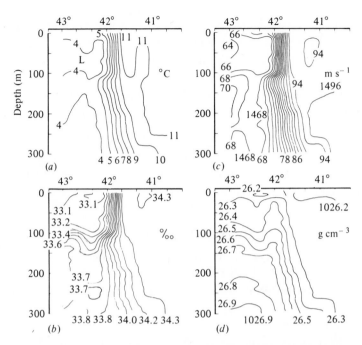

Fig. 1.11. Sections of (a) potential temperature, (b) salinity, (c) sound velocity and (d) potential density across the subarctic front at 168° E, occupied on 3 and 4 April 1971. The sections can be compared with the global section (Fig. 1.4). (From Roden, 1975.)

relatively low in biological activity, and blue in color (blue is the desert color of the sea). Sound velocity is relatively high. The boundaries of the gyres are marked by remarkably pronounced gradients in all these variables, and the gradients may be further concentrated into fronts. *Frontogenesis* is largely influenced by the wind field, and depends on the cross-frontal convergence of transport in a region of strong thermohaline gradients.

Roden (1975) has sketched the main oceanic fronts in the North Pacific (Fig. 1.10), and these bear an obvious relation to the circulation (Fig. 1.8). Fig. 1.11 shows a detailed section across the subarctic front at the boundary between the subtropical and subpolar gyres. Typical differences (southward) across the front are substantial: temperature $+8\,°C$, salinity $+1\,‰$, sound velocity $+28\,\mathrm{m\,s^{-1}}$, density $-0.2\times10^{-3}\,\mathrm{g\,cm^{-3}}$, current $0.4\,\mathrm{m\,s^{-1}}$. (The density jump is relatively small because of partial compensation between temperature and salinity.) In Figs. 1.10 and 1.11 the front is roughly 50 km wide. But there is a finestructure, and occasional jumps by a few degrees over a ship length have been observed. Dynamically, the density discontinuity $\delta\rho$ across a front is balanced by the Coriolis force, but this requires a (small) inclination m of the frontal surface according to

$$\delta\rho \, g \tan m = -2\Omega \sin \phi \, \delta(\rho u). \qquad (1.4.7)$$

Equation (1.4.7) is known as 'Margules' law' but was previously derived by Witte in 1878.

1.5. The surface mixed layer

The very upper layer of the ocean tends to be relatively uniform in temperature and salinity during the colder and windier part of the year, hence the term 'mixed layer'. During the calmer periods and those of warming, the upper layer is not well mixed, but weakly stratified in temperature.

The thickness of the mixed layer is typically from 10 to 100 m. At the bottom, the mixed layer can be terminated by a virtually discontinuous increase in the gradients of temperature and salinity. Within this 'thermocline' the buoyancy frequency n may exceed 10 cph; within the mixed layer n is generally less than 1 cph.

It is well established that the mixed layer is deepened by strong winds. In the presence of a persistent surface heat flux the erosion of gradients is arrested at some depth. With slackened winds and continuing heat flux the gradients are re-established. The dynamics of the mixed layer involves the processes of momentum, heat and salt flux.

Quantitative models have been attempted only recently (Pollard, Rhines and Thompson, 1973; Niiler, 1975). Following Niiler, let $u_* = (\tau/\rho)^{\frac{1}{2}}$ designate the friction velocity associated with the shear flow just beneath the surface due to wind stress τ. Then, following the onset of a strong wind, the mixed layer quickly deepens, and within a time $t = \pi/\omega_i$ (half the inertial period) reaches a depth

$$h_1 \sim 2u_*(n_0\omega_i)^{-\frac{1}{2}} \qquad (1.5.1)$$

where n_0 is the buoyancy frequency just beneath the mixed layer.

Subsequently, the deepening takes place at a diminishing rate, and is ultimately arrested at a depth

$$h_2 \sim 2Kqu_*^3/\dot{q} \qquad (1.5.2)$$

where q is the fraction of the turbulent energy generated in the surface wave zone that is available for mixing, $\dot{q}$ is the surface heat flux,

$$K = \rho C_P/ag \approx 5 \text{ cal s}^2 \text{ cm}^{-4} \qquad (1.5.3)$$

with a the coefficient of thermal expansion. Taking $\tau = 4 \text{ dynes cm}^{-2}$, $\rho = 1 \text{ g cm}^{-3}$, $u_* = 2 \text{ cm s}^{-1}$, $\dot{q} = 10^{-2} \text{ cal cm}^{-2} \text{ s}^{-1}$, $n_0 = 1.75 \times 10^{-2} \text{ s}^{-1}$ (10 cph), $\omega_i = 7.3 \times 10^{-5} \text{ s}^{-1}$, yields $h_1 = 35$ m, $h_2 = 80q$ m.

The Arctic surface layer, particularly during ice formation, represents a unique problem, and one of particular interest to acousticians. (For recent work see Foster, 1975.)

1.6. The canonical sound structure

In § 1.1, we found

$$C^{-1} \partial_z C = Gn^2(z) - \gamma_A. \qquad (1.6.1)$$

In shallow water the n^2 term dominates, and C increases upwards; in deep water $n^2 \to 0$ and the velocity increases downwards (with the rate γ_A). At the axis of the sound channel $\partial_z C = 0$, hence by (1.1.8) and (1.1.9)

$$n(z_1 = -h_1) \equiv n_1 = (\gamma_A/G)^{\frac{1}{2}} = 2.14 \times 10^{-3} s_1^{-\frac{1}{2}} \text{ s}^{-1}. \qquad (1.6.2)$$

So far, these relations are quite general, with G and n unspecified functions of depth. In the subsequent discussion of sound transmission it is very helpful to refer for illustration to a *canonical model*. For this purpose we return to the simplest case of an exponential ocean:

$$n = n_0 e^{z/B}, \quad r = b \, \partial_z S/a \, \partial_z T = \text{constant}, \qquad (1.6.3)$$

where n_0 is interpreted as a surface-extrapolated value. We then ignore the situations near the surface, at high latitudes and the Equator, and in the regions of the Western Boundary Currents. But in the remainder of the oceans, the conditions (1.6.3) constitute the simplest model that is not in conflict with the elemental facts of ocean structure. The severest remaining shortcoming is the neglect of local variations in $r(z)$ associated with intrusions of water masses.

From (1.6.2) we find that the depth of the sound axis is

$$-z_1 = h_1 = B \log (n_0/n_1) = B(0.89 + \tfrac{1}{2}\log s) \qquad (1.6.4)$$

where we have used $n_0 = 5.2 \times 10^{-3}\,\mathrm{s}^{-1}$, $\gamma_A = 1.14 \times 10^{-2}\,\mathrm{km}^{-1}$, $\alpha/a = 24.5$. Typically observed values are $-z_1 = 0.7$ to 1.5 km. Geographic variations in the sound axis are associated with the temperature dependence of n_1 through the a-parameter, and with the salinity dependence of $s(r)$. We shall use the parameters $-z_1 = B = 1$ km in much of the following discussion; this corresponds to $s = 1.25$, $r = +0.19$.

The canonical sound-speed gradients can be written

$$C^{-1}\,\partial_z C = \gamma_A(n^2 - n_1^2)/n_1^2, \quad C^{-1}\,\partial_z C_P = \gamma_A n^2/n_1^2.$$
$$(1.6.5)$$

In terms of a dimensionless distance η above the sound axis, the sound-speed profile is given by

$$C = C_1[1 + \varepsilon(e^\eta - \eta - 1)], \quad \varepsilon = \tfrac{1}{2}B\gamma_A = 5.7 \times 10^{-3},$$
$$\eta = (z - z_1)/\tfrac{1}{2}B. \qquad (1.6.6)$$

The coefficient ε is readily interpreted as the fractional adiabatic velocity increase over a scale depth. Equation (1.6.6) provides a reasonable description of an oceanic sound channel (see Fig. 2 in Munk, 1974), given in terms of physical constants of seawater and the stratification parameters n_0, B, and r.

For adiabatic vertical displacements ζ, the sound speed perturbation is given by

$$\mu \equiv \delta C/C = \zeta \cdot C^{-1}\,\partial_z C_P. \qquad (1.6.7)$$

The mean-square fluctuations are

$$\langle \mu^2 \rangle = \langle \zeta^2 \rangle (C^{-1}\,\partial_z C_P)^2. \qquad (1.6.8)$$

The potential sound-speed gradient is given by (1.1.9) for constant r, and for internal waves $\langle \zeta^2 \rangle$ varies as n^{-1}. Therefore

$$\langle \mu^2 \rangle = \langle \zeta_0^2 \rangle G^2 n_0 n^3 \equiv \langle \mu_0^2 \rangle (n/n_0)^3. \qquad (1.6.9)$$

The increase of vertical displacements with depth is more than offset by the decrease in the potential velocity gradient, and the sound-speed fluctuations diminish downwards as $n^{\frac{3}{2}}$.

If the additional assumption of the canonical profile is made,

$$\langle \mu^2 \rangle = \langle \mu_0^2 \rangle \, e^{3z/B}. \tag{1.6.10}$$

The canonical model with $z_1 = -B$ will serve for illustration in the following chapters. But in closing this narrative on ocean structure we emphasize once again the large geographical variations of the World Ocean, and the need for using concurrently measured local $C(z)$ and $n(z)$ in the interpretation of acoustical experiments.

CHAPTER 2

PLANETARY WAVES AND EDDIES

The classical oceanographer's aim was to measure the density field once and for all, so that ocean currents could then be deduced by the geostrophic method and printed on sailing charts and in textbooks. When the *Meteor* and *Altair* anchor stations in the 1930s revealed the presence of internal waves, these were regarded a nuisance, a high-frequency noise obstructing the production of current charts.

This idea of a steady circulation plus internal-wave noise came into difficulties when certain oceanographic stations were re-occupied and revealed changed conditions that could not always be dismissed as internal-wave noise or observational error. Gradually, the view developed that there were subinertial, quasi-geostrophic perturbations of the mean flow (much as in the atmosphere). Since the internal-wave equations have no wave-like solutions for $\omega < \omega_i$, Rossby adjusted the theory so that they do: by allowing for the latitudinal dependence of ω_i. The resulting planetary (or Rossby) waves do in fact exhibit some of the traits of ocean and atmosphere perturbations, such as a westward phase velocity, and have been, and are, a useful tool in the analysis of subinertial oscillations.

The ARIES measurements (Crease, 1962) gave the first *direct* evidence of midwater motion with subinertial fluctuations. Neutrally-buoyant Swallow floats were tracked acoustically, and revealed a variable structure with kinetic energy exceeding that of the mean motion by two orders of magnitude! A decade later the POLYGON (Brekhovskikh *et al.*, 1971) and MODE (Robinson, 1975) expeditions were mounted in a massive effort to map the subinertial variable flow field. We now know that a typical flow field in the upper ocean corresponds much more nearly to 1 ± 10 cm s^{-1} than 10 ± 1 cm s^{-1}, and this has far-reaching consequences.

The characteristic flow velocity U of order 10 cm s^{-1} is rather larger than typical phase velocities $C(k)$, and this suggests that interacting eddies with $\omega = Uk$ are a more useful representation than the dispersion $\omega = C(k)k$ of linear superimposed wave motion. Rhines (1975) has concluded that subinertial flow in the ocean is at the transition between wave-like and eddy-like motion, so that both viewpoints can contribute to our understanding.

2.1. Planetary waves

Unit vectors are $\hat{\mathbf{x}}$ east, $\hat{\mathbf{y}}$ north, $\hat{\mathbf{z}}$ up. In this section we use the traditional Rossby notation

$$\mathbf{f}_0 = \hat{\mathbf{z}}\omega_i, \quad \mathbf{f} = \mathbf{f}_0 + \boldsymbol{\beta}y. \tag{2.1.1}$$

In the internal-wave treatment we consider f as constant (f-plane dynamics). Here we permit f to vary linearly with distance northward. This generalization to the so-called β-plane leads to remarkably different dynamics, and seems to include most of the results of a true spherical geometry.

Neglecting nonlinear vertical velocity terms, the equations of motion in the β-plane are written

$$\partial_t\mathbf{u} + (\mathbf{u} \cdot \nabla)\mathbf{u} + \mathbf{f}\times\mathbf{u} + \nabla p = 0 \tag{2.1.2}$$

$$\partial_t w = -\partial_z p + b \tag{2.1.3}$$

where $\mathbf{u} = u\hat{\mathbf{x}} + v\hat{\mathbf{y}}$, $\mathbf{w} = w\hat{\mathbf{z}}$, $\nabla = \hat{\mathbf{x}}\,\partial_x + \hat{\mathbf{y}}\,\partial_y$, $\rho_0 p$ is pressure. $b = g(\rho_0 - \rho)/\rho_0$ is buoyancy, subject to

$$\partial_t b + n^2(z)w = 0. \tag{2.1.4}$$

For $\omega \ll f$, the term $\partial_t w$ is negligible, and we can combine (2.1.3) and (2.1.4) into

$$\partial_{tz}p + n^2 w = 0. \tag{2.1.5}$$

Now expand $\mathbf{u} = \mathbf{u}_0 + \mathbf{u}_1$, etc; for the zeroth-order solution, we take the geostrophic law

$$\mathbf{f}_0 \times \mathbf{u}_0 + \nabla p_0 = 0. \tag{2.1.6}$$

Equation (2.1.6) is consistent with a stream function ψ_0, such that

$$\mathbf{u}_0 = \hat{\mathbf{z}}\times\nabla\psi_0, \quad f_0\psi_0 = p_0. \tag{2.1.7}$$

The curl of (2.1.2) can now be put in the form

$$\mathrm{D}_t^0\,(\nabla^2\psi_0 + f_0^2\,\partial_z n^{-2}\,\partial_z\psi_0 + f) = 0 \tag{2.1.8}$$

where $D_t^0 = \partial_t + (\mathbf{u}_0 \cdot \nabla)$, and in particular $D_t^0 f = \beta\, \partial_x \psi_0$. Equation (2.1.8) describes the conservation of geostrophic potential vorticity along nearly horizontal particle trajectories.

The bottom boundary condition is

$$\mathbf{u} \cdot \nabla h + w = 0 \quad \text{at } z = -h(x, y).$$

For simplicity, we shall take a uniform north–south slope $\alpha = -\partial_y h$, which gives, using (2.1.5) and (2.1.7),

$$f_0 n^{-2} \partial_{tz} \psi_0 = -\alpha\, \partial_x \psi_0. \tag{2.1.9}$$

For purposes of illustration, take n constant, linearize by replacing D_t^0 by ∂_t and use the separable solution

$$\psi = \exp i(kx + ly - \omega t) \cos k_V z \tag{2.1.10}$$

which gives $\psi_z = 0$ at the surface (rigid lid). Substitution of (2.1.10) into (2.1.8) and (2.1.9) gives

$$\omega = -\beta k/(k^2 + l^2 + k_V^2 f_0^2 / n^2) \tag{2.1.11}$$

$$\theta \tan \theta = -\alpha k h n^2/(f_0 \omega) = +A(\theta^2 + \Theta^2) \tag{2.1.12}$$

where we have written for brevity $\theta = k_V h$, $A = \alpha f_0/\beta h$ and $\Theta^2 = (k^2 + l^2)h^2 n^2/f_0^2$. Following Rhines (1976), we now consider several asymptotic cases.

(i) *Barotropic waves.* This implies small vertical shear. Equation (2.1.12) then yields

$$\theta^2 = [A/(1 - A)]\Theta^2 \ll 1, \tag{2.1.13}$$

$$\omega = -k\beta(1 + A)/(k^2 + l^2). \tag{2.1.14}$$

For real k_V, $A = \alpha f_0/\beta h$ must be between 0 and 1. This requires that the magnitude of the slope be very small, less than the ratio of ocean depth to Earth's radius, i.e. $\alpha < 10^{-3}$. The classical Rossby wave is for $A = 0$, and corresponds to a westward (negative) phase velocity ω/k (2.1.14). Unless the slopes are exceedingly small ($A \ll 1$, $\alpha < 10^{-4}$), the barotropic condition $\theta \ll 1$ requires that $\Theta \ll 1$, e.g., wavelengths $2\pi(k^2 + l^2)^{-\frac{1}{2}}$ exceeding $2\pi n h/f_0 \approx 300$ km (see Fig. 2.1).

(ii) *Bottom-trapped waves.* For $A < 0$ or $A > 1$ we replace $\cos \theta$ by $\cosh \theta$ in (2.1.10) and this yields real values for θ, and leads again to the dispersion (2.1.14). For large $|A|$ these are the fast bottom-trapped waves discovered by Rhines.

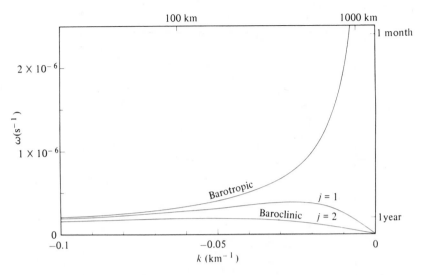

Fig. 2.1. Dispersion of the barotropic and lowest-order baroclinic waves for constant depth ($\alpha = 0$) and for $l = 0$.

(iii) *Baroclinic waves.* These have one or more vertical wavelengths in the water column. For vanishingly small slopes one has $\theta \approx j\pi$ from (2.1.12) and so

$$\omega = -\beta k/[k^2 + l^2 + (j\pi f_0/hn)^2], \quad j = 1, 2, \ldots \quad (2.1.15)$$

The maximum frequency is

$$\omega_m = -\beta/2k_m, \quad k_m^2 = l^2 + (j\pi f_0/hn)^2. \quad (2.1.16)$$

The most rapidly varying waves correspond to $l = 0$, $j = 1$, and these have a period of about a half year and a horizontal scale of $k_m^{-1} \approx 40$ km. For much larger scales, $k, l \ll \pi f_0/hn$,

$$\omega = -\beta(hn/\pi f_0)^2 k. \quad (2.1.17)$$

The corresponding expression for very large slopes is

$$\omega = -4\beta(hn/\pi f_0)^2 k. \quad (2.1.18)$$

Equations (2.1.17) and (2.1.18) give nondispersive, westward traveling modes. For a realistic density stratification, which is strongest in the upper ocean, the energy of baroclinic waves is confined to the upper ocean, and it may be appropriate to relate baroclinic waves to 'thermocline eddies'.

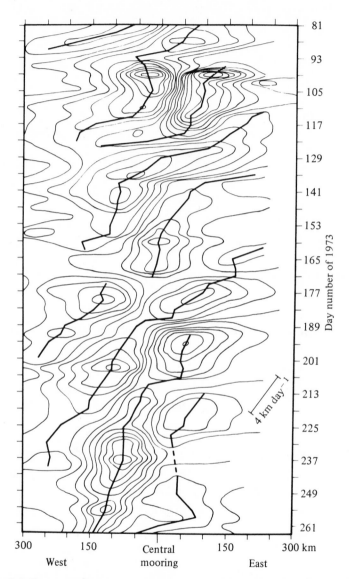

Fig. 2.2. East–west distance versus time plot of stream functions at a depth of 1500 m, along 28° N, centered at 69° 40′ W. Based on an objective analysis of SOFAR floats. A westward phase velocity of 4 km day^{-1} is indicated by the slope of the line.

Rhines (1976) has reviewed empirical evidence for the existence of these various modes. Streamlines based on SOFAR float tracks show definite evidence of westward phase velocity (Fig. 2.2). If these are barotropic Rossby waves, then the measured phase velocity of 5 km day^{-1} implies a horizontal scale of $k^{-1} = 50$ km, and a time scale $\omega^{-1} = 10$ days. The *difference* in dynamic height across the main thermocline exhibits a westward migration of about 2 km day^{-1}. The baroclinic Rossby wave (2.1.17) yields a westward migration of 1.5 km day^{-1}.

Effect of planetary waves on sound velocity

The planetary waves can affect sound velocity in two ways: by vertical displacement of the water structure, and by the flow field itself. The vertical displacement $\zeta = \int w \, dt$ follows from (2.1.5):

$$\zeta = -n^{-2} \, \partial_z p.$$

We have previously shown ((1.1.9), (1.6.5) and (1.6.7)) that

$$\delta C = (\partial_z C_P)\zeta = \gamma_A (n/n_1)^2 C\zeta = Gn^2 C\zeta. \qquad (2.1.19)$$

From (2.1.6), $\mathbf{f} \times \mathbf{u} = -\nabla p$. Thus, for solutions

$$p = p_0 \exp\left[i(kx + ly - \omega t)\right] \cos k_V z,$$

we have

$$u_H \sim f^{-1} k_H p_0, \quad \zeta \sim n^{-2} k_V p_0,$$

and so

$$u_H/\delta C = u_H/(Gn^2\zeta C) = (GfC)^{-1}(k_H/k_V) \approx 3(k_H/k_V) \qquad (2.1.20)$$

for $G = 3.13$ s m^{-1}, $f = 7.3 \times 10^{-5}$ s^{-1}, $C = 1500$ m s^{-1}. For barotropic waves we have shown (2.1.13) that k_H/k_V goes from ∞ for a constant depth, to 1 for a steeply tilting bottom, and so the flow effect is dominant. For baroclinic waves, we may select the condition $l = 0$, $k_H = j\pi f_0/hn$ for which ω is a maximum (2.1.15), and this would make $k_H/k_V = k_H/j\pi/h = f_0/n \approx 0.05$. Here the effect of vertical displacement is dominant.

The effect of Rossby waves on multipath sound propagation in an ocean with a bilinear sound profile has been considered by Baer and Jacobsen (1975).

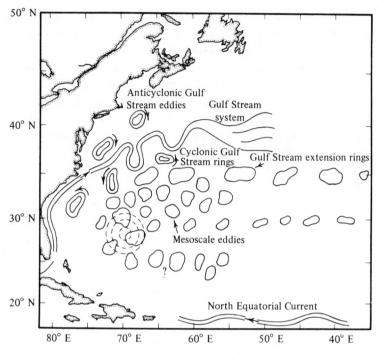

Fig. 2.3. A schematic representation of the variable ocean structure (Joint Polymode Organizing Committee, 1975).

2.2. Mesoscale

However strong the temptation to consider the ocean to be filled with classical linear waves, we must keep in mind that the mesoscale eddies '... are sufficiently strong to blow the tops off barotropic waves trying to propagate through them' (Rhines, 1976). Fig. 2.3 is a sketch of the flow field based on MODE and related measurements. The kinetic energy of the 'mesoscale eddies' at 1500 m depth diminishes from 9 cm^2 s^{-2} very near the Gulf Stream to 3 cm^2 s^{-2} at a distance of 500 km to the east. The energy level depends also on topographic features (such as the Mid-Atlantic Ridge) and on general bottom roughness. Since their documentation during POLYGON and MODE, mesoscale eddies have turned up in the Pacific Ocean and elsewhere. Most of the ocean's kinetic energy is associated with mesoscale eddies.

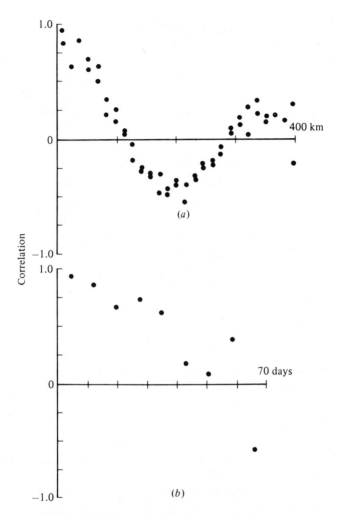

Fig. 2.4. (*a*) Spatial, and (*b*) temporal correlations of dynamic height across the main thermocline (Joint Polymode Organizing Committee, 1975).

Typical length and time scales are 100 km and 2 months respectively (Fig. 2.4). The correlation space scale is closely related to the internal Rossby radius of deformation

$$R_0 = C_0/\omega_i \qquad (2.2.1)$$

where C_0 is the phase velocity in the absence of rotation. Since C_0 is of order n_0B, we have $R = Bn_0/\omega_i$. (A typical wavelength is $4R_0$.)

Mesoscale currents diminish from 10 cm s^{-1} near the surface to 3 cm s^{-1} below the main thermocline. Take $u = 5 \text{ cm s}^{-1}$ at 1 km depth, at which depth the temperature anomaly for a typical mesoscale eddy is $\delta T = 1 \,°C$. The corresponding departure in sound velocity is $\delta C = a\delta T = 5 \text{ m s}^{-1}$, or 100 times the value of u. Alternatively, referring to (2.1.20), we get $u/\delta C = 10^{-2}$ for $k_H/k_V = 1/400$.

2.3. Geostrophic turbulence

Is there any hope that the low-frequency ocean variability can be described by a universal spectrum? Probably not, on account of the many parameters that affect variability, such as meteorological effects, distance from boundaries, and bottom roughness. But a certain general understanding has recently developed from oceanic and numerical experiments, and this could provide some guide in the context of this monograph. We refer the reader to the detailed discussion by Rhines (1976).

Geostrophic flow approximates two-dimensional turbulence. This is generally simpler, but quite different, from three-dimensional turbulence. In particular, the cascade of energy density $\mathscr{E}(k)$ is towards large scales (small k), as a result of an integral constraint on the squared relative vorticity, or enstrophy $k^2\mathscr{E}(k)$ (which itself has a flux towards small scales). For oceanic energy levels Rhines has demonstrated that energy may travel faster through wavenumber space than through physical space (important interactions occur in less than a wave period). The inability of geostrophic turbulence to cascade energy to small scales is a fundamental factor in preserving the relatively sharp boundaries of major ocean features, in contrast to the fuzzy structure of laboratory turbulence.

When the β-effect is included, the flux of baroclinic energy (currents with shear) converges on the Rossby deformation scale C_0/f from both larger and smaller scales. Subsequently, the eddies above and below the thermocline become nonlinearly coupled, and this would lead eventually to a barotropic state without vertical shear were it not for the opposing effect of topography and coastal boundaries.

The flux of barotropic energy towards larger scales is arrested at the Rossby wave scale, with the generation of a linear wave field of wavenumber $k = (\beta/2U)^{\frac{1}{2}}$ which is consistent with (2.1.13) for $U = \omega/k$ and $\alpha = 0$. In all, there appears to be some preference for a state of transition between waves and geostrophic turbulence, and this may be the reason why wave solutions seem to provide a useful guide in the study of oceans even though the underlying assumptions are grossly violated.

CHAPTER 3

LINEAR INTERNAL WAVES

Measurements of temperature and salinity fluctuations have given a picture of the orders-of-magnitude involved in internal-wave motions in the ocean. The following section (§ 3.1) describes some of the quantities of interest and their magnitudes, so that linear internal-wave theory can be developed from the equations of motion in suitable approximations (§§ 3.2 and 3.3). In §§3.4 and 3.5 the detailed spectrum of internal waves derived from temperature and salinity measurements is described. The final section (§3.6) summarizes the sound-speed fluctuations induced by internal waves.

3.1. Observed ocean fluctuations

Fluctuations of temperature and water velocity due to internal waves have been extensively observed. For the range of periods between a few minutes and one day, we show in Table 3.1 the order of magnitude of the fluctuations as a function of depth. Let

$\zeta \equiv$ displacement of an isodensity surface,

$v_z \equiv$ vertical component of water-particle velocity,

$v_x \equiv$ horizontal component of water-particle velocity,

$\delta T \equiv$ fluctuation in temperature at a fixed depth,

$\delta S \equiv$ fluctuation in salinity at a fixed depth,

$\delta\rho/\rho \equiv$ fractional fluctuation in density at a fixed depth, and

$\delta C/C \equiv$ fractional fluctuation in sound speed at a fixed depth.

The values in Table 3.1 are the result of an overall fit to many measurements of temperature and salinity, using the model for internal waves that is discussed in the next section. The values of $\delta C/C$ have been deduced from the measurements of temperature and salinity fluctuations, using the known equation of state for sea water.

Table 3.1. *Orders of magnitude of fluctuations in oceanic variables. Only contributions with periods in the range between a few minutes and one day are included. The primary source of these fluctuations is internal waves. The table entries are approximate rms values, and have been computed using the internal-wave spectrum discussed in this chapter, with the relations between temperature, salinity, density, and sound speed given by (1.1.4) to (1.1.9) with $r = 0.2$.*

Depth	ζ (m)	v_z (cm s^{-1})	v_x (cm s^{-1})	δT (°C)	δS (‰)	$(\delta\rho/\rho)$	$(\delta C/C)$	v_x/C
Near surface	7.3	0.5	4.7	0.14	5×10^{-3}	20×10^{-6}	5×10^{-4}	3×10^{-5}
1 km (near sound axis)	12.0	0.5	2.9	0.04	1×10^{-3}	4×10^{-6}	1×10^{-4}	2×10^{-5}
3–4 km (near bottom)	41.0	0.5	0.8	8×10^{-4}	3×10^{-5}	0.1×10^{-6}	3×10^{-6}	5×10^{-6}

The quantity v_x/C in Table 3.1 is a measure of sound-speed fluctuations due to internal-wave water-particle velocities acting through the Doppler effect. Under most circumstances, sound-speed fluctuations produced by temperature and salinity changes dominate the effects of internal-wave particle velocities, except at great depths, where v_x/C becomes comparable to $\delta C/C$. However, experiments involving reciprocal sound transmission between two points can isolate the particle velocity terms (Munk and Zachariasen, 1977).

Two other quantities are important in characterizing oceanic fluctuations; they are the horizontal and vertical wavelengths involved in the motions. If a picture were taken of an isodensity surface at an instant of time, it would have undulations due to internal waves (and other effects). A Fourier decomposition of this function yields the contribution of different wavelength intervals to the structure. The definition of a 'typical' wavelength is always difficult because it depends on how an average is taken. However, without being precise here (we remedy this in § 3.4) we may define two quantities, λ_V and λ_H, the vertical and horizontal wavelength of an ocean fluctuation. Clearly λ_V cannot be larger than 1 km, since λ_V larger than that would represent fluctuations of wavelength larger than the depth of the ocean. It is an important fact about the oceans that typically $\lambda_V \ll \lambda_H$; oceanic fluctuations are not isotropic.

Wavelength values characterizing the fluctuations listed in Table 3.1 are given in Table 3.2.

3.2. Equations for internal-wave motion

Basic equations

We begin with the assumption of an inviscid, incompressible fluid which, however, is allowed to have variable density. Subsequently, in order to deal with sound, we relax the condition of incompressibility. The variation of density is caused by variation of temperature, and is of the order of 0.1 per cent from the top to the bottom of the ocean. We also assume that a given water element has an unchangeable density during its motion (that is, heat conduction is negligible).

Table 3.2. *Values of fluctuation wavelength.*

	Minimum	Peak of spectrum	Maximum
λ_V	10 m	100 m	1000 m
λ_H	0.5 km	5 km	50 km

Let ρ be the density of fluid, $\mathbf{u}$ be the velocity of a fluid element with components (u, v, w), p be the pressure and $\hat{\mathbf{k}}$ be a unit vector in the vertical (z) direction. The convective derivative is defined as

$$D_t = \partial_t + \mathbf{u} \cdot \nabla, \tag{3.2.1}$$

We may write the equations of motion as follows:

$$\rho D_t \mathbf{u} = -\nabla p + \rho g \hat{\mathbf{k}}, \tag{3.2.2}$$

$$D_t \rho + \rho \nabla \cdot \mathbf{u} = 0, \tag{3.2.3}$$

$$\nabla \cdot \mathbf{u} = 0, \tag{3.2.4}$$

where (3.2.2) is Newton's second law (called Euler's equation in this context); (3.2.3) is the equation of continuity (or nondiffusivity of heat in our case); and (3.2.4) is the requirement of incompressibility.

Consider first the static case ($\mathbf{u} = 0$). Then the equations reduce to the hydrostatic pressure equation

$$\nabla p_0 = \rho_0 g \hat{\mathbf{k}} \tag{3.2.5}$$

where ρ_0 is the equilibrium density distribution, and is a function of z, not a constant.

Define the time varying parts of the density (ρ') and pressure (p') by

$$p = p_0 + p', \quad \rho = \rho_0 + \rho', \tag{3.2.6}$$

and (3.2.2) reduces to

$$(\rho_0 + \rho') D_t \mathbf{u} = -\nabla p' + \rho' g \hat{\mathbf{k}}. \tag{3.2.7}$$

At this point most derivations of internal waves introduce the Boussinesq approximation, which is to replace $\rho_0 + \rho'$ in the inertial term on the left-hand side of (3.2.7) by a constant. (That is, not by ρ_0 which is a function of z, but by a constant.) However, we do not make the Boussinesq approximation, first because it is unnecessary in our case, and second because our subsequent derivation of the equations of sound propagation in the presence of internal waves is not possible if we make the Boussinesq approximation. (See Chapter 5). We show how the Boussinesq approximation falls out of our derivation at the end of this section.

Linearization

We now assume that the motions are infinitesimal so that we may linearize the above equations. We neglect all second-order terms. We have:

$$\rho_0 \, \partial_t \mathbf{u}_H = -\nabla_H p', \qquad (3.2.8)$$

$$\rho \, \partial_t w = -\partial_z p' + \rho' g, \qquad (3.2.9)$$

$$\nabla_H \cdot \mathbf{u}_H + \partial_z w = 0, \qquad (3.2.10)$$

$$\partial_t \rho + w \, \partial_z \rho_0 = 0, \qquad (3.2.11)$$

where the index H denotes horizontal.

Solution

Plane-wave solutions of these equations have the form

$$w = W(z) \, e^{i(kx - \omega t)}, \quad \mathbf{u}_H = \mathbf{U}_H(z) \, e^{i(kx - \omega t)}, \quad \text{etc.} \quad (3.2.12)$$

and eliminating variables other than w from (3.2.8) to (3.2.11) yields

$$\partial_{zz} W + \left(\frac{n^2(z)}{\omega^2} - 1 \right) k^2 W = \frac{n^2(z)}{g} \partial_z W. \qquad (3.2.13)$$

For internal waves in the ocean we can now show that the right-hand side of (3.2.13) is negligible. Taking k_V as a vertical wavenumber for internal-wave fluctuations we may estimate the ratio of the right-hand side to one of the terms on the left-hand side,

$$[n^2(z)/g] \, \partial_z W (\partial_{zz} W)^{-1} \approx n^2/g k_V \leqslant 4 \times 10^{-4}, \qquad (3.2.14)$$

where we have taken the smallest reasonable value of k_V, 1 km^{-1} (see § 3.1).

Thus the right-hand side may always be neglected. Following the derivation back, we may note that the right-hand side is just the term neglected in the Boussinesq approximation.

We may now write:

$$\partial_{zz} W + \left(\frac{n^2(z)}{\omega^2} - 1 \right) k^2 W = 0. \tag{3.2.15}$$

Modifying this equation to account for the rotation of the earth we find (Eckart, 1960):

$$\partial_{zz} W + \left(\frac{n^2(z) - \omega^2}{\omega^2 - \omega_i^2} \right) k^2 W = 0, \tag{3.2.16}$$

where ω_i is the inertial frequency of (1.1.1).

By integration (3.2.16) is also the equation for the displacement, ζ, of internal waves. We will henceforth use (3.2.16) because it properly prevents internal-wave frequencies below ω_i. However we shall ignore the anisotropy introduced into the horizontal wavefunctions for internal waves created by the rotation of the Earth.

To solve (3.2.16) we must find the nature of the boundary conditions. At the bottom of the ocean ($z = z_B$), no vertical motion can occur, so that $W(z_B) = 0$. The surface boundary condition is more complicated in principle, but in practice one may also set $W(0) = 0$. The quantitative justification is given in Phillips (1966). A simple physical argument may also be given to convince the reader of the validity of this surface condition, as follows:

The restoring force of internal-wave motion is reduced compared with surface-wave motion by a factor $\Delta\rho/\rho \sim 10^{-3}$ due to the buoyancy. Furthermore internal waves move considerably more water mass than do surface waves. As a result, internal waves travel roughly one hundred times slower than surface waves of the same horizontal wavelength. Viewing the boundary as a surface–internalwave coupling, this coupling is clearly far off resonance, and so the amplitude of the driven surface wave will be very small.

For a given value of k, with the above surface and bottom boundary conditions, and with the exponential $n(z)$ function introduced in Chapter 1, we may solve (3.2.16) either numerically or analytically (Garrett and Munk, 1972). The two boundary conditions create an eigenvalue problem, so that only discrete

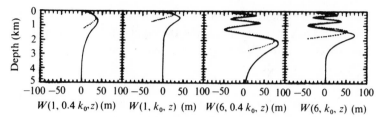

Fig. 3.1. Examples of normalized eigenfunctions of internal-wave displacement. The value 0.5 cycles km^{-1} was used for k_0. The dotted curves are obtained from a WKBJ approximation expressed in (3.3.3).

values of ω lead to a solution. Hence for each value of k there is a series of modes $W(j, k, z)$ with corresponding frequencies $\omega(j, k)$. Fig. 3.1 shows a number of examples of $W(j, k, z)$, and Fig. 3.2 shows the dispersion relations; that is $\omega(j, k)$. Fig. 3.3 shows the horizontal phase speed for internal waves.

Validity of linearization

We may now investigate whether our linearization of the equations was justified in view of the known amplitudes of internal waves described in § 3.1.

An example will suffice: we compare the two terms of the convective derivative;

$$\frac{\mathbf{u} \cdot \nabla}{\partial_t} \sim \frac{V_z k_V}{\omega} \sim \frac{(10^{-3}\,\mathrm{m\,s^{-1}})(10^{-2}\,\mathrm{m^{-1}})}{5 \times 10^{-3}\,\mathrm{s^{-1}}} \sim 10^{-3}, \quad (3.2.17)$$

so that the linearization is well justified.

Orthogonality, normalization and the energy integral

We have derived a set of basis wavefunctions $W(j, k, z)$ for the vertical component of the water-particle velocity. By integration, $W(j, k, z)$ is also the set of wavefunctions for the vertical displacement, ζ, of a water-particle from its equilibrium position.

The orthogonality condition for these functions can be derived by using the differential equation (3.2.16).
We obtain, for $j \neq j'$,

$$\int_{z_B}^{0} W(j, k, z) W(j', k, z)[n^2(z) - \omega_i^2]\,\mathrm{d}z = 0. \quad (3.2.18)$$

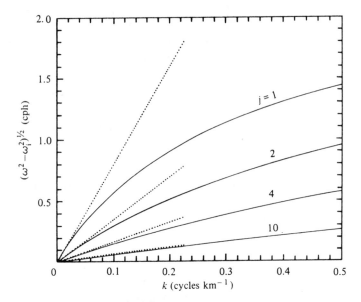

Fig. 3.2. Dispersion relations for internal waves. The dotted lines are the approximate linear forms obtained from an improvement of (3.5.2) where j is replaced by $j - 0.25$.

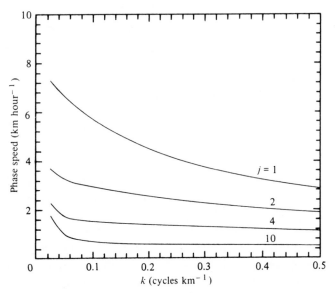

Fig. 3.3. Horizontal phase speed (ω/k) of internal waves.

Equation (3.2.18) suggests an orthonormality condition which we shall use to obtain an energy normalization. The energy of an internal wave is rather difficult to write down because of the complication of a rotating coordinate system with spherical equipotential surfaces. However, we can at least indicate the physics involved by neglecting the rotation of the earth. Then we may write the potential energy density of an internal wave as:

$$E_P = (A\tau)^{-1} \int \tfrac{1}{2}\rho_0 n^2(z)|\zeta|^2 \, dz \, dx \, dy \, dt, \qquad (3.2.19)$$

where E_P is the potential energy per unit area, A is the area over which the x–y integral is done, and τ is the time over which the t integral is done. Integrating over x, y, and t yields:

$$E_P = \int_{z_B}^{0} \tfrac{1}{2}\rho_0 n^2(z) W^2(k, j, z) \, dz. \qquad (3.2.20)$$

It can be shown straightforwardly in this case that the kinetic energy density is equal to the potential energy density so that

$$E = 2E_P. \qquad (3.2.21)$$

Approximating ρ_0 by a constant, ρ_1, we may then combine (3.2.19) to (3.2.21) to give an energy orthonormality condition in the absence of rotation:

$$\rho_1 \int_{z_B}^{0} n^2(z) W(k, j, z) W(k, j', z) \, dz = \delta_{jj'}. \qquad (3.2.22)$$

A careful inclusion of the rotation of the earth modifies this condition to our final orthonormality condition:

$$D_1 \int_{z_B}^{0} [n^2(z) - \omega_i^2] W(k, j, z) W(k, j', z) \, dz = \delta_{jj'},$$

$$(3.2.23)$$

where D_1 is different from ρ_1 by terms of order ω_i/ω. We shall use $D_1 = \rho_1$, which results in some distortion of the energy spectrum near the inertial frequency.

It is important to note that GM72 ignored the effect of rotation on the energy integral by leaving the ω_i^2 term out of the integral; thus their spectrum is also distorted near the inertial frequency compared with a true energy distribution. It is probably not worthwhile at the present time to attempt an estimate of the exact energy spectrum because the derivation will depend on the planetary-wave description, which in turn depends on the exact shape of the ocean basin used, as well as on various approximations. Since the spectrum of internal waves is derived from experimental observations, the distortion in the energy integrals is compensated by a distortion in the spectrum in order to fit the data. Hence in a practical sense the distortion has little effect on predictions of various fluctuations.

3.3. Approximation to the wavefunctions $W(k, j, z)$

Equation (3.2.16) with the exponential $n(z)$ given by (1.6.3) is, in fact, Bessel's equation with solutions:

$$W(k, j, z) = AQ_a(\beta), \quad a = kB(1 - \omega_i^2/\omega^2)^{-\frac{1}{2}}, \quad \beta = an(z)/\omega,$$
(3.3.1)

where

$$Q_a(\beta) = J_a(\beta) - [J_a(\beta_h)/Y_a(\beta_h)]Y_a(\beta),$$
(3.3.2)

where J_a and Y_a are the Bessel functions of the first and second kind. A is a normalization constant chosen to satisfy (3.2.23). The quantity h is the depth of the ocean ($h = -z_B$). If $h \gg B \ln(n_0/\omega_i)$ then the second term in $Q_a(\beta)$ is negligible and it is sufficient to use the Bessel function of the first kind. Unfortunately this inequality does not hold for the real ocean, so that precise calculations require Bessel functions of the second kind.

The discrete values of j result from matching this general solution to the boundary condition at the surface; i.e. $Q_a[an(0)/\omega] = 0$.

Several of the derivations in Part III of this book will use an approximation introduced by GM72 in which:

$$W(k, j, z) = [\tfrac{1}{2}\rho_1 n_0 Bn(z)]^{-\frac{1}{2}} \cos[\beta - (a\pi/2) - (\pi/4)] \text{ for } \omega < n(z),$$

$$= 0 \quad \text{for } \omega > n(z).$$
(3.3.3)

This approximation is compared with the exact functions in Fig. 3.1. It is seen that significant errors, of the order of 10 per cent, will result even from the use of averages of (3.3.3) in certain cases, although (3.3.3) does hold the advantage of simplicity.

3.4. The spectrum of internal waves

Fluctuations in the vertical structure of temperature and salinity were discovered by Petterson, Helland-Hansen and Nansen soon after the turn of the century. Since that time there has been a vast literature on the subject (over 500 references were compiled by J. Roberts, 1973) consisting mostly of reports of temperature and current fluctuations recorded by moored instruments, and of a few horizontal temperature profiles recorded by instruments towed behind ships. In the past three years, the technology of continuous vertical profiling of currents with freely dropped instruments has been developed, providing additional information. A three-dimensional trimooring (IWEX) was installed in 1973 off the American east coast, and we may expect some very useful additional results.

On the basis of this myriad of observations, Garrett and Munk have contrived successive models (Garrett and Munk, 1972, 1975) of internal-wave spectra. They placed particular emphasis on multiple recordings, separated vertically on the same mooring or horizontally on neighbouring moorings, which had shown that fluctuations of frequencies as low as 1 cph were uncorrelated for vertical separations exceeding a few hundred meters, and for horizontal separations exceeding a few kilometers. These coherences were interpreted as a measure of reciprocal bandwidth since, for separations larger than the reciprocal bandwidth, different wavenumbers interfere destructively, and coherence is lost. The following conclusions were reached:

(i) Observations can be reconciled with the dispersion law and the wavefunctions of linear internal-wave theory.

(ii) Records compiled from towed instruments are insensitive to the ship's course, and records compiled from moored instruments are similar for the two velocity components, thus indicating some degree of horizontal isotropy; the evidence is certainly incompatible with internal waves propagating along narrow horizontal beams.

(iii) Coherences are incompatible with a model consisting of just the gravest one or two vertical modes (except at tidal frequencies). The GM72 model had equal contributions from modes 1 to 20, and none beyond mode 20. Recent measurements by Cairns and Williams (1976), however, are consistent with a mode weighting according to $(j^2 + j_*^2)^{-1}$ with $j_* = 3$. (Apparently some of the coherence loss in the earlier experiments can be attributed to instrument noise.)

Numerous observations, taken over the years at many depths off the west and east American coasts, off Hawaii, Bermuda and Gibraltar, in the Bay of Biscay and the Mediterranean, agree to within an order of magnitude. This suggests some universality in the internal-wave spectrum, perhaps due to saturation effects such as those limiting surface waves of high frequency.

One may also note that the total energy density in internal waves, as derived by Garrett and Munk (1975), is 0.4 joules cm^{-2}, integrated from the surface to the bottom of the ocean. This value is approximately equal to the average energy density for surface waves, perhaps suggesting that the coupling between surface waves and internal waves is strong enough to set up some sort of equilibrium between them.

Details of the spectrum

The spectrum of internal waves was originally described by Garrett and Munk in a space characterized by horizontal wavenumber k and frequency ω, treating vertical structure with what was essentially a WKBJ approximation so that the spectrum in ω was continuous rather than discrete. However, we present a spectrum in the more mathematically straightforward k, j space, and show in § 3.5 how to translate this spectrum into other spaces such as that of Garrett and Munk.

The displacement, ζ, of an isodensity surface may be represented as a superposition of internal waves in the following way:

$$\zeta = \sum_{k_1, k_2, j} G(k_1, k_2, j) W(k, j, z) e^{i[k_1 x + k_2 y - \omega(k, j)t]} \qquad (3.4.1)$$

where $k = (k_1^2 + k_2^2)^{\frac{1}{2}}$, and (k_1, k_2) are the components of the wavenumber vector in the (x, y) directions. The summation symbolizes two-dimensional integration in horizontal wavenumber space, and summation over mode number j. The quantities $G(k_1, k_2, j)$ are complex Gaussian random variables with zero mean, thus representing the internal-wave amplitudes as a statistical mixture.

A quantity of interest, to be derived from this spectrum, is the energy density as a function of k and j averaged over all of space.

(Garrett and Munk start their derivations from an assumed expression for the energy density.) Accounting for both kinetic and potential energy, we may write the total energy density

$$E = \rho_1[n^2(z) - \omega_i^2]|\zeta|^2. \tag{3.4.2}$$

Averaging over x, y, and t, and integrating over z to find the energy density per unit area, we can see that from the orthonormality condition (3.2.23)

$$E = \sum_{k_1, k_2, j} |G(k_1, k_2, j)|^2, \tag{3.4.3}$$

so that the spectral energy density is given by

$$E(k_1, k_2, j) = |G(k_1, k_2, j)|^2. \tag{3.4.4}$$

The spectrum derived from oceanographic observations will be expressed in terms of the expectation value of $|G|^2$, where it will be remembered that G is a complex random variable with zero mean (or expectation value). It is convenient to first use the assumption of horizontal isotropy to define a new quantity

$$G(k, j) = (2\pi k)^{-\frac{1}{2}} G(k_1, k_2, j) \tag{3.4.5}$$

so that

$$\int |G(k, j)|^2 \, dk = \int\int |G(k_1, k_2, j)|^2 \, dk_1 \, dk_2 \tag{3.4.6}$$

and the energy density in k, j space is

$$E(k, j) = |G(k, j)|^2. \tag{3.4.7}$$

Then the observed ocean fluctuations can be represented by

$$\langle |G(k, j)|^2 \rangle = E_0 H(j) B(k, j), \tag{3.4.8}$$

$$H(j) = N(j^2 + j_*^2)^{-p/2}, \tag{3.4.9}$$

$$B(k, j) = (2/\pi) k_j k^2 (k^2 + k_j^2)^{-2}, \tag{3.4.10}$$

$$k_j = (\pi/B)(\omega_i/n_0)j. \tag{3.4.11}$$

The functions $H(j)$ and $B(k, j)$ are normalized such that

$$\sum_j H(j) = 1, \tag{3.4.12}$$

$$\int B(k, j) \, dk = 1. \tag{3.4.13}$$

Equations (3.4.1), and (3.4.8) to (3.4.13) describe the spectrum completely except for the numerical values of the modal power law,

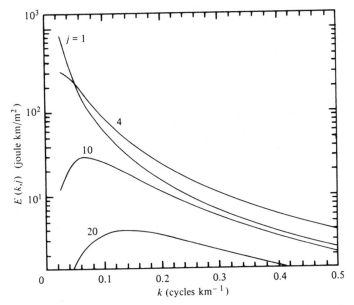

Fig. 3.4. Energy spectrum of internal waves as a function of mode number j and horizontal wavenumber k.

p, the average energy, E_0, and the characteristic mode number j_*. When numerical values are needed, we shall use $p = 2$, $E_0 = 0.4$ joule cm^{-2}, and $j_* = 3$. (For $p = 2$, N can be approximated by $2j_*/\pi$.)

Fig. 3.4 shows the contributions of the various modes to the energy. We note that the lower j numbers have most of the energy and that the energy is also concentrated near low horizontal wavenumbers. Nevertheless, Part III of this book shows that the high j numbers have an important effect on sound scattering because of their rapid spatial variation.

3.5. Equivalent spectra

It will be useful, both for historical reasons and for analytical treatment, to express the internal-wave spectrum in two other spaces. Frequency–mode-number space (ω, j) was used in Garrett and Munk, 1975, and frequency–horizontal-wavenumber space (ω, k) was used in Garrett and Munk (1972).

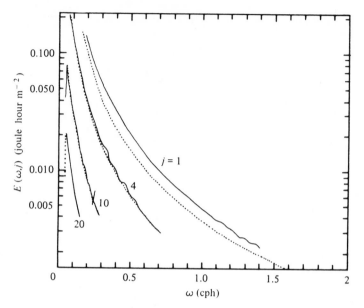

Fig. 3.5. Energy spectrum of internal waves as a function of mode number j and frequency ω. This spectrum has been obtained from $E(k, j)$ via the exact dispersion relation (solid lines) and via the approximate dispersion relation (dotted lines).

Frequency–mode-number space

For each j mode a value of frequency can be obtained from the horizontal wavenumber k by use of the dispersion relation $\omega(k, j)$. Then the energy densities can be related by:

$$E(\omega, j) = E(k, j)[d\omega(k, j)/dk]^{-1}. \qquad (3.5.1)$$

For low ω ($\omega < 0.5n_0$) the dispersion relation may be approximated (GM72)

$$k = (j\pi/Bn_0)(\omega^2 - \omega_i^2)^{\frac{1}{2}}. \qquad (3.5.2)$$

Hence

$$E(\omega, j) = (j\pi/Bn_0)[\omega^2/(\omega^2 - \omega_i^2)]^{\frac{1}{2}}E(k, j), \qquad (3.5.3)$$

where (3.5.2) must be used to replace k in $E(k, j)$. Fig. 3.5 shows the exact $E(\omega, j)$ derived from (3.5.1) and (3.4.7) to (3.4.13), as well as the approximate values obtained using (3.5.2).

Wavenumber–frequency continuum space

In this section we translate the discrete j spectrum into a continuous frequency spectrum. Again the dispersion relation is required, but this time its important dependence is on j:

$$E(k, \omega) = E(k, j)[d\omega(k, j)/dj]^{-1}. \qquad (3.5.4)$$

Using the approximate dispersion relation (3.5.2) we find:

$$E(k, \omega) \approx (kB/\pi)n_0(\omega^2 - \omega_i^2)^{-3/2}E(k, j), \qquad (3.5.5)$$

where the dispersion relation must be used to replace j by a function of ω and k. In both cases $E(k, \omega)$ is normalized so that

$$\sum_j \int E(k, j)\,dk = \iint E(k, \omega)\,dk\,d\omega. \qquad (3.5.6)$$

Wavevector space

The discrete mode number, j, can be transformed to a continuum variable, that may be considered as a vertical wavenumber, analogous to the two horizontal wavenumbers k_1 and k_2. This may be done only *locally* in depth, since at different depths a given mode, j, corresponds to different vertical wavelengths. Letting k_V be the wavenumber conjugate to vertical displacement of an isodensity surface, we have:

$$k_V^2 = k^2(n^2 - \omega^2)/(\omega^2 - \omega_i^2), \qquad (3.5.7)$$

so that

$$E(\mathbf{k}) = E(k_1, k_2, \omega)(d\omega/dk_V). \qquad (3.5.8)$$

3.6. The sound-speed correlation function

It has been shown in Chapter 1 that the variation in sound speed is related, to a good approximation, to the displacement due to internal waves by the relation:

$$\mu(\mathbf{x}, t) \equiv \delta C/C = (C^{-1} \partial_z C_P)\zeta = Gn^2(z)\zeta. \qquad (3.6.1)$$

Hence, assuming knowledge of G and $n(z)$, any result for the quantity ζ can be transformed immediately into a result for the sound-speed fluctuation, μ, by multiplication by the factor $Gn^2(z)$. Fig. 3.6 shows wavefunctions for sound-speed deviation, obtained

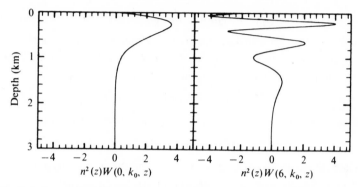

Fig. 3.6. Examples of internal-wave eigenfunctions of sound-speed deviation from equilibrium. The value 0.5 cycles km^{-1} was used for k_0.

by multiplying $W(j, k, z)$ by $n^2(z)$. We note that while the displacement may be large below a depth of 1 km, the sound-speed fluctuation is not, due to the small gradient of potential sound speed (the n^2 factor).

We may then calculate the correlation function ρ defined by:

$$\rho[\mathbf{x}-\mathbf{x}', t-t', (\mathbf{x}+\mathbf{x}')/2] \equiv \langle \mu(\mathbf{x}, t)\mu(\mathbf{x}', t')\rangle. \qquad (3.6.2)$$

For internal waves ρ will depend on $(\mathbf{x}+\mathbf{x}')/2$ only through the average vertical position $\bar{z} = (z + z')/2$, and the $t-t'$ variable is not independent of the $\mathbf{x}+\mathbf{x}'$ variable; they are connected through the dispersion relation for internal waves. One can then show that

$$\rho(\mathbf{x}-\mathbf{x}', \bar{z}) = [G^2 n^4(\bar{z})]\langle \zeta(\mathbf{x})\zeta(\mathbf{x}')\rangle. \qquad (3.6.3)$$

Defining a wavevector $\mathbf{k}$, conjugate to $\mathbf{x}-\mathbf{x}'$, we define the Fourier transform of ρ as F;

$$F(\mathbf{k}, \bar{z}) \equiv (2\pi)^{-3} \int d^3y \, e^{-i\mathbf{k}\cdot\mathbf{y}} \rho(\mathbf{y}, \bar{z}). \qquad (3.6.4)$$

Using (3.4.1) it can then be shown that F is expressible in terms of the energy spectrum of internal waves.

However, the path from (3.4.8) to (3.4.13), which define the energy spectrum of internal waves, to useful expressions for ρ or F is a long one, and has in fact not been traveled except with the approximation (3.3.3) to the wavefunctions $W(k, j, z)$. These approximations yield forms (GM75) for F remembering that through the dispersion relation the vector $\mathbf{k}$ may be replaced by the variable sets $\{\omega, j\}$ or $\{k, j\}$;

$$F(\omega, j, \bar{z}) = \langle \mu_0^2 \rangle [n(\bar{z})/n_0]^3 (4/\pi) \omega_i (\omega^2 - \omega_i^2)^{\frac{1}{2}} \omega^{-3} H(j), \qquad (3.6.5)$$

$$F(k, j, \bar{z}) = \langle \mu_0^2 \rangle [n(\bar{z})/n_0]^3 4(\omega_i/n_0) B^{-1} k^2 j H(j)$$
$$\times \{k^2 + [j\pi\omega_i/(n_0 B)]^2\}^{-2} \qquad (3.6.6)$$

$$F(\mathbf{k}, \bar{z}) = \langle \mu_0^2 \rangle [n(\bar{z})/n_0]^2 (2/\pi^3)(\omega_i/n_0) k |k_V| k_V^*$$
$$\times [k^2 + (\omega_i k_V/n)^2]^{-2} [k_V^2 + k_V^{*2}]^{-1} \qquad (3.6.7)$$

where $k_V^* = j_*(\pi/B)(n/n_0)$ and where $\langle \mu_0^2 \rangle$ is the variance of the sound-speed fluctuations extrapolated to the ocean surface, with a typical value of 2.5×10^{-7}. The approximations that have been made have resulted in a simple $\bar{z}$ dependence; in general of course the $\bar{z}$ dependence would be intimately intertwined with the other variables (e.g. $\{k, j\}$). The restriction $\omega_i < \omega < n(\bar{z})$ applies to (3.6.5).

From (3.6.6) the following important characteristics of internal-wave-induced sound-speed fluctuations can be deduced:

(i) Horizontal scales are long compared with vertical scales, typically by a factor of ten to a hundred.

(ii) The fluctuations are largest near the surface and decrease rapidly with depth [approximately like $n^3(\bar{z})$].

(iii) As a function of vertical wavenumber, k_V, the spectrum falls like a power of k_V, for $k_{max} > k_V > k_{min}$ where $2\pi/k_{min} \gtrsim 10^2$ m and $2\pi/k_{max} \lesssim 1$ m. This conclusion follows from identifying $k_V \approx j\pi n/(n_0 B)$. The particular power-law exponent is called p for a one-dimensional spectral function.

PART II

INTRODUCTION TO SOUND TRANSMISSION IN THE OCEAN

Officer (1958) has carefully detailed the two main treatments of the wave equation for underwater sound when the sound-speed profile is a simple deterministic function of depth. These treatments yield solutions in terms of rays on the one hand, or normal modes on the other. In this book our detailed theories are applied to cases where the water is deep enough and the sound frequencies are high enough that the ray picture has physical usefulness. We shall therefore use the language of ray theory and geometrical optics throughout, in preference to normal modes; making clear at each stage where we are taking the geometrical-optics result, and where we are using extensions and corrections to geometrical optics. In particular Chapter 5 describes the path from the full wave equation to the geometrical-optics approximation with two very useful intersections on the way; the parabolic approximation and the path-integral formulation.

CHAPTER 4

THE OCEAN SOUND CHANNEL

The most striking feature of ray propagation in the ocean is the sound channel, discovered by Ewing and Worzel (1948). The sound-speed profile has a minimum at a depth that varies with geographical location, but is typically about 1 km. Rays tend to bend toward regions of smaller sound speed and are thereby *channeled* toward the depth of minimum speed, called the sound axis. As a result refracted rays may extend for thousands of kilometers in range, never touching surface or bottom.

As discussed in § 1.6, the typical sound-speed profile is a result of the competition between the stratification of the ocean (in temperature and salinity) which tends to give a sound-speed decrease with depth, and the adiabatic pressure gradient, which tends to give an increase with depth. The stratification dominates in the upper regions and the pressure gradient dominates in the lower depths. Quantitatively this competition is expressed in (1.6.1). Assuming an exponential decrease in $n(z)$ with depth, Munk (1974) has *derived a canonical sound-speed profile* in terms of physical constants of seawater and the stratification parameters n_0 and B (see (1.6.3)).

This chapter will describe the ray structure for the canonical profile presented in Chapter 1, using the geometrical-optics approximation.

4.1. Rays in the sound channel

In the geometrical-optics approximation, the field transmitted from a source can be described completely in terms of rays, whose characteristics are determined by the sound speed as a function of position. The frequency of sound, σ, enters only to the extent that the phase along a ray varies in a simple manner according to

$\phi = \int q \, ds$ where ds is the element of path length and q is the local wavenumber. We define $q = \sigma/C$, where $C(z)$ is the local sound speed.

The general equation of ray optics for a ray in a depth-dependent sound-speed profile may be expressed as

$$\partial_{xx}z = -C^{-1}\partial_z C[C_0 \sec^4\theta/C] \tag{4.1.1}$$

where θ is the angle of the ray with the horizontal x coordinate; $\partial_x z = \tan\theta$ and C_0 is an arbitrary reference sound-speed. The above equation is a generalized Snell's law, derived from the eikonal equation. In the ocean case the term in parentheses is close to unity (see below). Equation (4.1.1) may be used to numerically integrate a ray from any initial starting depth and angle. Since $\partial_z C$ is positive above the sound axis and negative below the sound axis, a general result of (4.1.1) is that rays always bend toward the sound axis. This is the principle of the *sound channel*.

Fig. 4.1 shows a set of rays from a source on the sound axis propagating to a range of 70 km. After a few tens of kilometers the sound field has a rather complicated structure. Several rays from a single source may arrive at a receiver from different directions, giving rise to *multipath*. Regions exist through which no rays pass (shadow zones). Envelopes to families of rays exist, called caustics. These effects have been extensively studied (e.g. Weinberg, 1975).

Because the sound speed $C(z)$ never varies by more than a few per cent from the reference C_0, fully refracted rays within the ocean volume never attain angles of more than ~15° from the horizontal, a fact important to the parabolic-equation approximation (see Chapter 5). In the ray approximation this is equivalent to setting $\tan\theta = \theta$; we may then replace (4.1.1) by

$$\partial_{xx}z = -C^{-1}\partial_z C. \tag{4.1.2}$$

Fig. 4.1 gives a qualitative understanding of sound-channel propagation; we now present some additional quantitative results that will be useful in subsequent chapters.

The ray shown in Fig. 4.2 has a radius of curvature, r, at the upper turning point, or apex, given by

$$r^{-1} \equiv |\partial_{xx}z| = C^{-1}\partial_z C = (2\varepsilon/B)(e^{\eta_a} - 1), \tag{4.1.3}$$

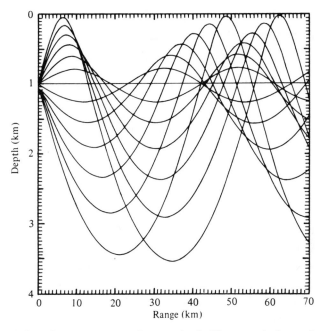

Fig. 4.1. Rays from a source on the sound axis. The canonical sound-speed profile (1.6.3) was assumed with $-z_1 = B = 1$ km.

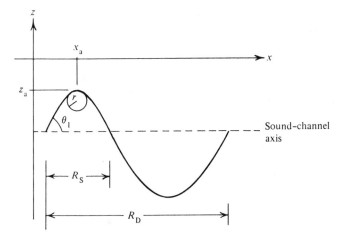

Fig. 4.2. Any ray in a range-independent environment is characterized by the parameters of its double loop.

where η_a is $2(z_a - z_1)/B$. If the apex is near the *ocean surface* we have approximately

$$r \approx [2\varepsilon(e^2 - 1)]^{-1}B = 13.7 \text{ km.}$$

The ranges indicated in Fig. 4.2 can be approximately expressed (Pedersen, 1968, and Munk, 1974) by

$$R_S = R_0[1 + (2\sqrt{2}/3\pi)\phi_a + (1/12)\phi_a^2 + \cdots] \quad (4.1.4)$$

$$R_D = 2R_0[1 + (1/12)\phi_a^2 + \cdots] \quad (4.1.5)$$

where

$$R_0 \equiv (\pi/2)\varepsilon^{-\frac{1}{2}}B = 20.8 \text{ km,} \quad (4.1.6)$$

and

$$\phi_a = \pm[e^{\eta_a} - \eta_a - 1]^{\frac{1}{2}}, \qu(4.1.7)$$

with the $-$ or $+$ sign taken for upper or lower loops respectively. We see that upper loops have shorter ranges, downward loops have longer ranges, as compared with R_0. The range of a double loop always exceeds that of the *axial double loop*, given by $2R_0$.

To a good approximation the range of a single loop varies linearly with the apex height (Munk, 1974)

$$\partial_{z_a}R_S \approx (R_0/B)(4\sqrt{2}/3\pi)(e^{\eta_a} - 1)(e^{\eta_a} - \eta_a - 1)^{-\frac{1}{2}} \quad (4.1.8)$$

which, for small η_a, is given by

$$\partial_{z_a}R_S \approx \tfrac{4}{3}(R_0/B) = 8.8. \quad (4.1.9)$$

Thus for every meter increase in apex height, the single loop range decreases by about nine meters.

Another quantity of interest is the acoustic path length of a double loop. We have (Munk, 1974)

$$S_D \approx R_D - (\varepsilon R_0/12)\phi_a^4{}^{\dagger}. \quad (4.1.10)$$

Thus the axial ray ($\phi_a = 0$) arrives last in a SOFAR transmission. We will require

$$\delta \equiv R_D(d^2S_D/dR_D^2) = -12\varepsilon[1 + (1/12)\phi_a^2 + \cdots] \quad (4.1.11)$$

[dagger] The equation for S_D in Munk and Zachariasen (1976) is incorrect.

Evaluating for the axis ray ($\phi_a = 0$) and for a surface-limited ray ($\phi_a = 2.1$) yields

$$\delta = -0.07 \quad \text{(flat ray)} \tag{4.1.12}$$

$$\delta = -0.09 \quad \text{(steep ray)} \tag{4.1.13}$$

4.2. Angle–depth diagrams

Most often a physical feeling for rays is obtained by plotting depth versus range for a group of rays (e.g. Fig. 4.1). However, plots of angle versus depth, although somewhat more abstract, have in many cases more powerful applications. Fig. 4.3 is the angle–depth diagram for the canonical profile.

The angle θ is measured from the horizontal with positive angle upward. Since the sound-speed profile is range independent, each ray forms a closed curve; that is, any given ray repeats the same angle and depth after some range. The ocean surface is at $z = 0$, and the bottom has been taken at 4 km depth.

As a ray travels in range, it traverses its curve in a counterclockwise fashion. Thus range may be indicated by distance along the curve. The starting point is arbitrary; here the starting point is taken at the sound axis with the ray pointing upward; that is, at the depth where the ray has the largest angle with the horizontal. The small tick marks are at 1 km intervals, and the large ones are at 10 km intervals. The ray is labeled by its largest angle. For example, the 8° ray starts at (8°, 1 km); at a range of 7.5 km it reaches its minimum depth (0°, 0.3 km); at a range of 14.9 km it crosses the sound axis (−8°, 1 km); at a range of 31 km it reaches its maximum depth (0°, 2.3 km); and at a range of 47.5 km returns to its starting position.

Rays that hit the surface or bottom can be treated in several ways; here they are assumed to undergo specular reflection. In that case the depth remains at zero (or 4 km) while the angle changes sign in a negligible range interval. Thus a ray starting at 18° reaches the surface at a range of 3.4 km where $\theta = 13°$. It appears at 3.4 km to continue with $\theta = -13°$. It is important to note that whatever the treatment of rays that hit the surface or bottom, these rays always remain *outside the shaded (fully refracted) area of the diagram*.

Even at this simple level the diagram is useful for answering such questions as: given a ray with 4° angle and 0.5 km depth, with what

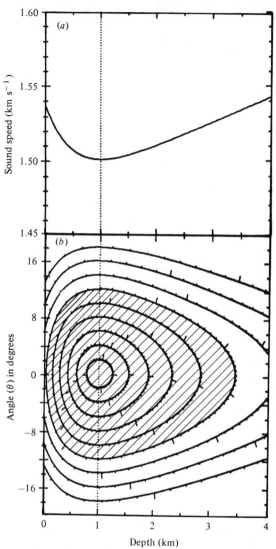

Fig. 4.3. Sound-speed profile and angle–depth diagram for the canonical profile. (a) Sound speed as a function of depth. The sound axis is at 1 km depth. Values of constants are; $-z_B = B = 1$ km; $\varepsilon = 0.0057$; $C_1 = 1.5$ km s^{-1}. (b) Angle–depth diagram. Angles are measured from the horizontal. Each curve represents a ray propagating in range. All curves may be thought of as starting from the sound axis and going counterclockwise from the positive angle indicated by the curved label. Each small tick mark on the curve represents a movement of 1 km in range; each large tick mark indicates 10 km in range. Rays that reflect from surface or bottom always remain outside the shaded (fully refracted) area. (From Flatté, 1976.)

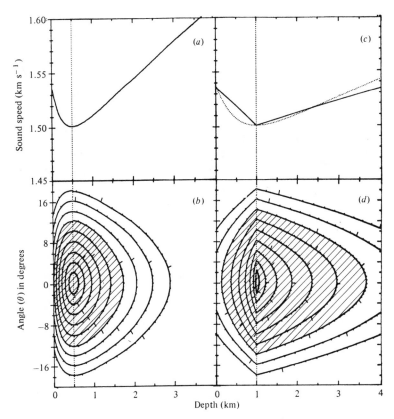

Fig. 4.4. Other sound-speed profiles. (a) and (b): profile and angle–depth diagram for $-z_B = B = 0.5$ km in (1.6.3). (c) and (d): profile and angle–depth diagram for a bilinear profile with $dC/dz = 0.035$ s^{-1} for $-z < 1$ km, and $dC/dz = 0.0114$ s^{-1} for $-z > 1$ km. The dashed curve in (c) shows the profile of Fig. 4.3 for comparison. (From Flatté, 1976.)

angle does the ray cross a depth of 1.8 km? (Answer, obtained by following near the 6° ray, is about 2°.) A fine grid on such a plot can yield accurate answers much more easily than plots of angle or depth versus range.

Fig. 4.4 shows angle–depth diagrams for two other sound-speed profiles; one with a sound axis at 500 m and another using a bilinear approximation to the profile used in Fig. 4.3.

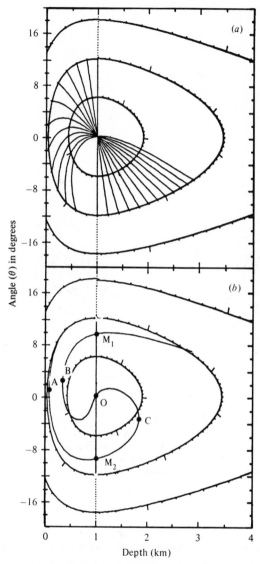

Fig. 4.5. Source lines for an axis source. (*a*) A source at depth 1 km emitting energy between ±12° is indicated by the vertical line. The subsequent ten lines show the source line propagating in range in one-kilometer steps to a range of 11 km. (*b*) The line is the same source as in (*a*) but propagated to a range of 100 km. Points M_1, O, and M_2 illustrate a multipath situation with three rays arriving at the same depth with different angles. Points A, B and C are caustics (tangents to a vertical line). (From Flatté, 1976.)

Source lines

A point source is indicated on the angle–depth diagram as a source line. As a bundle of rays from this point source traverses range, each point on the source line will trace out its ray path, distorting the source line into a new shape. Thus all rays from the source are followed simultaneously.

Fig. 4.5(a) shows a source line for a source at 1 km depth (the sound axis) emitting sound rays at angles between ±12°. The original source line is vertical. The source line is shown at 1 km intervals up to a range of 11 km, where its shape has changed considerably. If this source line is followed to a range of 100 km, the curve in Fig. 4.5(b) is obtained. The line has been curled up, but of course it can never cross itself, as that would imply that two rays of different angles from the point source arrived at a later range with the same depth and angle.

A more complete discussion of angle–depth diagrams, including treatments of multipath, caustics, obstacles, and Liouville's theorem, is contained in Flatté (1976). An approximate angle–depth diagram has been discussed by Cox (1977).

CHAPTER 5

THE WAVE EQUATION

Treatments of the mathematics of underwater sound often begin with the three-dimensional wave equation. However, the wave equation is an approximation to the general equations of fluid motion. In a case where one is interested, not in the gross features of wave propagation, but in small fluctuations in wave propagation, it should be shown that the terms neglected in deriving the wave equation have in fact a negligible effect on fluctuations. In addition, one of the important tools used in this book is the parabolic approximation to the wave equation. The relation of this approximation to the approximations necessary to derive the wave equation in the first place is of interest. Finally, since the ray picture is such an important element in understanding underwater sound and its fluctuations, the emergence of geometrical optics from the wave equation should be very clear in the reader's mind.

The following sections describe quantitatively how the wave equation (§ 5.1) and the parabolic approximation to the wave equation (§ 5.2) are derived from the general equations of motion. The ray picture (§ 5.4) is then developed by use of the path-integral formulation (§ 5.3), a device utilized in later chapters on fluctuations.

5.1. Fundamental approximations

The basic equations are the equation of continuity (3.2.3) and Euler's equation (3.2.2). The general problem involves sound waves and water motion. For the sake of concreteness we will consider the water motion as resulting from internal waves. Thus, we must keep clearly in mind that we are dealing with the superposition of two wave motions: (1) internal waves whose motions are incompressible and rotational; and (2) sound waves whose motions are irrotational and compressional. Both these motions will be considered as small, so that second-order terms may be neglected. The fluid velocity **u** and the density ρ may be expressed as:

$$\mathbf{u} = \mathbf{u}_i + \mathbf{u}_s, \tag{5.1.1}$$

$$\rho = (\rho_0 + \rho')(1 + s), \tag{5.1.2}$$

where $\mathbf{u}_i$ and $\mathbf{u}_s$ are the fluid velocities due to internal waves and sound waves respectively, ρ_0 is the equilibrium density distribution (a function of z), ρ' is the density variation due to internal waves, and s is the fractional density change due to the sound wave. Incompressibility of internal waves and the irrotational character of sound imply that we may write to first order:

$$\nabla \cdot \mathbf{u}_i = 0, \tag{5.1.3}$$

$$\mathbf{u}_s = \nabla\phi, \tag{5.1.4}$$

where ϕ is a scalar field.

Neglecting second-order terms, the equation of continuity in terms of these variables is:

$$\partial_t\rho' + w\,\partial_z\rho_0 + \rho_0\,\partial_t s + \partial_z\phi\,\partial_z\rho_0 + \rho_0\nabla^2\phi = 0, \tag{5.1.5}$$

where w is the z component of $\mathbf{u}_i$.

The first two terms of (5.1.5) are internal-wave terms, with characteristic internal-wave frequencies, while the last three terms are sound-wave terms with frequencies that are generally larger by orders of magnitude. Therefore the equation separates into (3.2.11) and:

$$\partial_t s + \partial_z\phi\rho_0^{-1}\,\partial_z\rho_0 + \nabla^2\phi = 0. \tag{5.1.6}$$

Treating Euler's equation (3.2.2), neglecting second-order terms and separating frequencies we can show that

$$\partial_{tt}\nabla\phi + \rho_0^{-1}\,\partial_t\nabla p_s - g\,\partial_t s\,\hat{\mathbf{k}} = 0, \tag{5.1.7}$$

where p_s is the part of the pressure due to the sound wave.

Some manipulation of derivatives yields

$$\nabla(\partial_{tt}\phi + \rho_0^{-1}\,\partial_t p_s) + \rho_0^{-1}\,\partial_t p_s(\nabla\rho_0/\rho_0) - g\,\partial_t s\hat{\mathbf{k}} = 0. \tag{5.1.8}$$

The speed of sound is defined through

$$\rho_0^{-1}\,\partial_t p_s = \rho_0^{-1}\,\partial_s p_s\,\partial_t s = C^2\,\partial_t s, \tag{5.1.9}$$

where C is the speed of sound. Combining (5.1.6), (5.1.8) and (5.1.9), we have:

$$\nabla(\partial_{tt}\phi - C^2\nabla^2\phi - C^2\,\partial_z\phi\,\rho_0^{-1}\partial_z\rho_0) + \partial_t s[C^2(\nabla\rho_0/\rho_0) - g\hat{\mathbf{k}}] = 0.$$
(5.1.10)

At this point we neglect all except the first two terms of the equation; such approximation will be justified below. Without loss of generality, the resulting equation is

$$\partial_{tt}\phi - C^2\nabla^2\phi = 0. \tag{5.1.11}$$

The key to our treatment is that we will now derive a reduced wave equation whose terms will then be compared with the neglected terms above.

5.2. The reduced wave equation and the parabolic approximation

Invoking the cylindrical symmetry of the equilibrium state, and the wave nature of the sound field, it will be useful to set

$$\phi = r^{-\frac{1}{2}}\,e^{i(q_0 r - \sigma t)}\psi(\mathbf{r}, t). \tag{5.2.1}$$

Neglecting terms that fall off in range faster than $r^{-\frac{1}{2}}$, (5.1.11) becomes:

$$(C_0/C)^2\sigma^{-2}\,\partial_{tt}\psi - (2i/\sigma)(C_0/C)^2\,\partial_t\psi - q_0^{-2}\nabla^2\psi$$
$$- (2i/q_0)\,\partial_r\psi + 2U\psi = 0, \tag{5.2.2}$$

where $C_0 = \sigma/q_0$ and

$$U = (C^2 - C_0^2)/(2C^2). \tag{5.2.3}$$

Expanding $\nabla^2\psi$ (neglecting $1/r$ terms), we have

$$\nabla^2\psi = \partial_{rr}\psi + r^{-2}\,\partial_{\phi\phi}\psi + \partial_{zz}\psi. \tag{5.2.4}$$

Now we are prepared to compare the sizes of various terms. Consider the reduced wavefunction ψ. Its variations can be characterized in an order-of-magnitude fashion by a horizontal wavenumber, k_H; a vertical wavenumber, k_V; and a frequency ω_a. Typical observed values of these quantities are 10^{-3} m^{-1}, 10^{-2} m^{-1}, and 10^{-2} s^{-1}. We know that U from the sound channel is $\sim 10^{-2}$, and from internal waves is $\sim 10^{-4}$.

Table 5.1 shows the order of magnitude of terms for a frequency $\sigma \sim 10^3$ s^{-1} and wavenumber $q_0 \sim 1$ m^{-1}. The first three terms of

Table 5.1. *Terms in the equation for the reduced wavefunction of sound amplitude, and their orders of magnitude.*

Term number	Term last appeared in equation	Expression	Order-of-magnitude expression	Order of magnitude
1	(5.2.2)	$U\psi$	U	10^{-4} to 10^{-2}
2	(5.2.2)	$q_0^{-1}\,\partial_r\psi$	k_H/q_0	10^{-3}
3	(5.2.2)	$q_0^{-2}\,\partial_{zz}\psi$	$(k_V/q_0)^2$	10^{-4}
4	(5.2.2)	$q_0^{-2}r^{-2}\,\partial_{\phi\phi}\psi$	$(k_H/q_0)^2$	10^{-6}
5	(5.2.2)	$q_0^{-2}\,\partial_{rr}\psi$	$(k_H/q_0)^2$	10^{-6}
6	(5.2.2)	$\sigma^{-1}\,\partial_t\psi$	ω_a/σ	10^{-5}
7	(5.2.2)	$\sigma^{-2}\,\partial_{tt}\psi$	$(\omega_a/\sigma)^2$	10^{-10}
8	(5.1.10)	$(C/\sigma)^2\rho_0^{-1}\partial_z\rho\,\partial_z\phi$	n_0^2/gq_0	10^{-8}
9	(5.1.10)	$\sigma^{-2}\int(\partial_t s)C^2(\nabla\rho_0/\rho_0)\,\mathrm{d}x$	n_0^2/gq_0	10^{-8}
10	(5.1.10)	$\sigma^{-2}\int(\partial_t s)g\,\mathrm{d}x$	q_0g/σ^2	10^{-5}

Table 5.1 are the largest. Term 4 is small; its neglect is justified if one is not interested in horizontal deflection of sound, but if the exchange of energy in the horizontal plane is to be studied, term 4 must be included. *Neglect of term 5 is called the parabolic approximation*, since its neglect changes an elliptic equation into a parabolic equation (see Leontovich and Fok, 1946; Tappert and Hardin, 1973, 1974; Tappert, 1974; Hardin and Tappert, 1973). It is valid presuming our estimates of k_V and k_H are valid, and basically depends on k_H/k_V being small. In addition it has had to be assumed that the sound energy travels at angles close to the horizontal in order to use the simple order-of-magnitude expressions for terms 4 and 5. Otherwise terms like $k_V \tan\theta/q_0$ would appear. Relative to the first three terms, the remaining terms (6–10) can be neglected, although term 6, which comes from the time variation of the internal-wave field, and term 10, which comes from the effect of gravity on the sound field, are rather large. It is of interest to note that starting with the three-dimensional Helmholtz equation corresponds to using terms 1–5. Inclusion of term 5 would be inconsistent with the neglect of terms 6 and 10.

Thus, for studying the fluctuation of sound due to internal waves, in a region of the ocean where the sound travels more or less horizontally, and for frequencies in the region of 100 Hz or higher, we use the *parabolic wave* equation, corresponding to terms 1–3;

$$-(1/2q_0^2)(\partial_{zz}\psi + r^{-2}\partial_{\phi\phi}\psi) + U\psi = (i/q_0)\partial_r\psi. \quad (5.2.5)$$

We now find it useful to redefine our variables into a Cartesian system: let $y = r\phi$ and $x = r$. Then (5.2.5) becomes:

$$-(1/2q_0^2)(\partial_{zz}\psi + \partial_{yy}\psi) + U\psi = (i/q_0)\partial_x\psi. \quad (5.2.6)$$

Since in general the dependence of U on y is considerably weaker than on z, the y dependence of ψ is weak compared with the z dependence, as discussed previously (terms 3 and 4 of Table 5.1).

5.3. Introduction to the path-integral formulation

We shall show that a solution of (5.2.6) can be found in an iterative form suitable for use on a large computer. We then show that this solution can be recast in terms of path integrals that are useful in the study of approximations to solutions of the wave equation.

Assume that U is a slowly varying function of range. Then the evolution of the wavefunction in range is given by:

$$\psi(x + d) = e^{i(A+B)d}\psi(x),$$
$$A = (1/2q_0)(\partial_{zz} + \partial_{yy}), \quad (5.3.1)$$
$$B = -q_0 U,$$

where d is an infinitesimal increment in range.

Neglecting the weak y dependence for simplicity, and expressing (5.3.1) in terms of Fourier transforms we find:

$$\psi(z, x + d) = F^{-1}\{e^{-ik^2d/2q_0}F[e^{-iq_0 Ud}\psi(z', x)]\} \quad (5.3.2)$$

where F is the Fourier transform from z-space to k-space. This equation is useful because of the fast-Fourier-transform technique so widely applied in digital computers at the present time. Taking d as a finite range step, (5.3.2) can be used to step a wavefunction through a range-dependent sound-speed profile. Hardin and Tappert (1973) noticed that with a finite step size it is useful to use

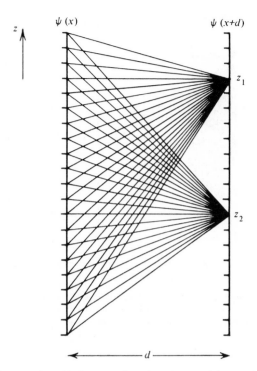

Fig. 5.1. Propagation of the wavefunction from $\psi(x)$ to $\psi(x+d)$. The calculations for $z = z_1$ or $z = z_2$ are illustrated.

instead the equation:

$$\psi(z, x+d) = e^{-\frac{1}{2}iq_0 Ud} F^{-1}\{e^{-i(k^2 d/2q_0)} F[e^{-\frac{1}{2}iq_0 Ud}\psi(z, x)]\}. \quad (5.3.3)$$

This 'split-step' algorithm is valid to a higher order in the small parameter d. (A computer solution usually involves many steps; the difference between implementing (5.3.2) or (5.3.3) occurs only at the first and last steps.)

The relation of the computer method to path integrals is developed by doing the k integral explicitly in (5.3.2). One then finds

$$\psi(z, x+d) = (q_0/2\pi id)^{\frac{1}{2}} \int e^{-iq_0 Ud}\psi(z', x) e^{+i(q_0/2d)(z-z')^2} dz'. \quad (5.3.4)$$

Fig. 5.1 illustrates a physical interpretation of (5.3.4). Assume the wavefunction ψ is known at x for all z'. Then $\psi(z, x+d)$ is deter-

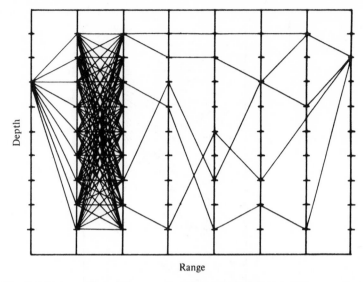

Fig. 5.2. Propagation of the wavefunction over a finite range by a series of short steps. For one step all possible paths are shown; for the other steps only representative ones have been drawn. The diagram illustrates both the path-integral method and the iterative computer solution.

mined by an integral of $\psi(z', x)$ over z', with each contribution requiring a phase advance. The phase advance is due to two effects: (1) the difference from nominal sound speed in the region of z'; and (2) the extra path length of the line connecting (z', x) to $(z, x + d)$ compared with the step d. The latter effect is often called the Green's function, or the propagator, for the field from (z', x) to $(z, x + d)$.

The propagation of a wave field over a finite range R by a series of short steps is illustrated in Fig. 5.2. If the grid spacings on this figure are considered infinitesimal, the operations correspond to an exact solution to the wave equation. If the grid spacings are considered finite, the operations correspond to a solution by digital computer using (5.3.3) or (5.3.4). In either case the illustrated operation is called the method of path integrals, because the operation corresponds to summing all possible paths from any point on the initial wave to a point where the wavefunction is desired. It is important to realize that 'all possible paths' does not mean paths of real sound rays, but rather all single-valued functions that can be drawn from the initial to the final point. The operation can be regarded as Huygens' principle repeatedly applied.

Rather than considering an initial wavefunction at all z', it will be useful to consider the case of a point source at (z', x'). Dividing $x - x'$ into n steps, the above discussion implies the following expression for the wavefunction, where the assumption of propagation at small angles to the horizontal is used:

$$\psi(\mathbf{x}) = (i/4q_0)^{\frac{1}{2}} \lim_{n \to \infty} \int \left(\prod_{j=1}^{n-1} dz_j \right) \left(\frac{q_0 n}{2\pi i (x - x')} \right)^{n/2}$$

$$\times \exp \left\{ iq_0 \frac{x - x'}{n} \sum_{j=1}^{n} \left[\frac{n^2}{2} \left(\frac{z_j - z_{j-1}}{x - x'} \right)^2 - U(x_j, z_j) \right] \right\}, \quad (5.3.5)$$

where $x_j = x' + (j/n)(x - x')$.

It has been mentioned that this expression can be visualized as a sum over all possible paths through the grid. If we let $n \to \infty$ we may write the wavefunction for a point source at $\mathbf{x}$ as

$$\psi(\mathbf{x}) = (i/4q_0)^{\frac{1}{2}} \int d(\text{paths}) \exp \left\{ iq_0 \int_{\text{path}} [\tfrac{1}{2}(\partial_x z_{\text{path}})^2 - U] \, dx \right\}, \quad (5.3.6)$$

where we identify the first two factors in (5.3.5) as the differential in path space, and the exponent as the sum of two integrals, one involving the geometry of the path and the other involving the sound-speed deviations.

It can be shown that a correct treatment of the weak y dependence that was neglected from (5.3.2) to (5.3.6) yields a generalization of (5.3.6):

$$\psi = (i/4q_0)^{\frac{1}{2}} \int d(\text{paths}) \exp [iq_0 S(\text{path})], \quad (5.3.7)$$

$$S(\text{path}) = \int_{\text{path}} \{ \tfrac{1}{2}(\Delta y/R)^2 + \tfrac{1}{2}[\partial_x z(x)]^2 - U \} \, dx \quad (5.3.8)$$

where Δy is the difference in the y coordinates of the end points of the path (separated by range R), $z(x)$ is the z coordinate of the path at range x, and in U the value of $y(x)$ is taken from the unperturbed ray.

5.4. Rays

When q_0 is very large, one may evaluate (5.3.5) by the method of stationary phase since the bulk of the integral results from paths that satisfy the equation

$$\tfrac{1}{2}[n/(x-x')]^2(z_{j+1}-2z_j+z_{j-1})+\partial_z U(x_j, z_j)=0. \quad (5.4.1)$$

Setting $n \to \infty$ we find

$$\partial_{xx}z(x)+\partial_z U(x, z(x))=0. \quad (5.4.2)$$

But (5.4.2) is just the equation for rays through a variable sound-speed field, from Fermat's principle and in the parabolic approximation (i.e., equation (4.1.2)). Thus rays come naturally out of the path-integral formulation through the stationary-phase approximation, an approximation that is better as q_0 becomes larger.

Evidently, the idea of wave propagation as an integral over paths shows that rays have a real, very specific meaning in the wave theory. For example one can ask: when are two nearby rays really distinct? The answer is simple: they have to be distinct stationary-phase points in the integral over paths. If this is to be the case, the phase of the integrand at one ray must differ from the phase at the other ray by at least unity. If t_1 and t_2 are the travel times along the two rays, they will be distinct if $|t_1-t_2|>1/\sigma$ where σ is the frequency of sound.

Actually, this is a sufficient but not a necessary condition. Two rays could be well separated but might accidentally have the same travel time. The precise condition is that there should be no way to deform one ray into the other without incurring a travel time change of at least $1/\sigma$ at some point in the deformation.

The wavefunction in the ray approximation

If the evaluation of (5.3.7) by stationary phase is valid, then the pressure at the receiver can be expressed as a sum over rays (if there are more than one), and each ray contributes

$$\psi(\mathbf{x}_1, \mathbf{x}_2) = K(\mathbf{x}_1, \mathbf{x}_2) \exp[iq_0 S(\mathbf{x}_1, \mathbf{x}_2)], \quad (5.4.3)$$

where

$$S(\mathbf{x}_1, \mathbf{x}_2)= \int_{\mathbf{x}_1}^{\mathbf{x}_2} \{\tfrac{1}{2}[(y_2-y_1)/R]^2+\tfrac{1}{2}[\partial_x z(x)]^2 - U\} \, dx. \quad (5.4.4)$$

The quantity S is called the *eikonal*; the normalization factor $K(\mathbf{x}_1, \mathbf{x}_2)$ depends on certain derivatives of $S(\mathbf{x}_1, \mathbf{x}_2)$, and is given more explicitly in Part IV of this book. The integral in (5.4.4) is understood to be evaluated along the *ray* from $\mathbf{x}_1$ to $\mathbf{x}_2$. Note that S is in fact the acoustic path length along the ray from $\mathbf{x}_1$ to $\mathbf{x}_2$ (with a constant value R subtracted), and S/C_0 is the travel time along the ray (with a constant R/C_0 subtracted).

Region of validity

The validity of the combination (5.4.3) and (5.4.4) can be expressed as a general condition as follows: consider the ray from $\mathbf{x}_1$ to $\mathbf{x}_2$, through the sound-speed field expressed by $U(\mathbf{x})$. Now consider any *path* (not necessarily a ray) deformed from the ray. The difference between the travel times for the ray and the path may be formed. If

$$q_0[S(\text{path}) - S(\text{ray})] > 1 \tag{5.4.5}$$

for all paths which are separated from the ray at any point by more than a scale size L of sound-speed fluctuations, then the stationary-phase evaluation is valid.

We may give an example of the application of (5.4.5) for the sound channel itself. The scale length L of the sound channel is ~ 1 km. It can be shown that in almost all geometries, if a path is deformed by more than 100 m, (5.4.5) is satisfied for ranges out to several thousand kilometers for wavelength less than 10 m. It is important to note that if the sound channel were not regular, but consisted of statistical inhomogeneities of scale size 1 km, then (5.4.5) would fail at a range $\sim (10^6/\lambda)$m. Hence the regularity of the sound channel is an important qualitative consideration in sound-fluctuation theory.

However, if U contains perturbations on the sound channel, such as internal waves, then the application of (5.4.5) becomes extremely complicated. Many treatments have been given for homogeneous isotropic perturbations without a background sound channel (see, for example, Tatarskii, 1971). The present work treats anisotropic, internal-wave perturbations on a background sound channel (see Parts III and IV of this book).

One last remark about solutions to the wave equation should be made here. If the sound-speed deviations U are small enough, then first-order perturbation theory can be used. This alternative to the ray method can be pursued either through the path-integral method, or more conventionally by starting from the wave equation (Chapter 10). Again, treatments have previously been given (particularly by Rytov, 1937) for isotropic perturbations in the absence of a sound channel. Our treatment considers the sound channel and anisotropic perturbations.

PART III

SOUND TRANSMISSION THROUGH A
FLUCTUATING OCEAN

Wave propagation through statistically homogeneous, isotropic fluctuations in an otherwise constant index of refraction has received much attention. The main impetus for work in this field has come from consideration of optical and radio-wave propagation through the atmosphere, although the problems of propagation through interplanetary plasma, and sound propagation in both the atmosphere and the oceans have had their influence. An historical review of treatments of the wave equation is not attempted here, but a few landmarks might be helpful. Pioneering work was done by Rytov (1937). In the 1950s progress considerably accelerated through a combination of theory and experiment. Chernov (1960) has presented a unified general treatment of weak-scattering methods. Salpeter (1967) and Cohen *et al.* (1967) have given a discussion of both weak- and strong-scattering regimes in the context of interplanetary scintillations. They have presented the different regimes in a parameter space that we generalize to the ocean transmission case in this Part (see also de Wolf, 1975). Tatarskii (1971) has given a complete treatment of the weak-scattering regime under the conditions of isotropic turbulence, and in addition has made some progress in the strong-scattering region using a Markov approximation (see also Beran, 1970).

The ocean sound channel (Chapter 4) and the anisotropy in the sound-speed fluctuations from internal waves are two important qualitative differences between the ocean and the atmosphere as far as wave propagation is concerned. (Atmospheric internal waves have not played an important role.) Thus the ocean is a layered medium with a random component. Ewing, Jardetzky and Press (1957) and Brekhovskikh (1960) have considered waves in layered media, but have given little consideration to the random component. Brekhovskikh *et al.* (1975) have briefly treated the random component from the point of view of isotropic turbulence; however in describing experimental results they have commented that internal-wave effects are probably dominant. Finally, Beran (1975) has reviewed some work he did with co-workers on including the background sound channel with isotropic turbulence, or treating the anisotropic case with no sound channel (this case was solved for a particular angle of incidence, and only if the sound wavelength was larger than the vertical outer scale of fluctuations, which in the case of internal waves would mean $\lambda > 200$ m).

Our treatment of the ocean medium includes the effects of anisotropy and the background sound channel, as well as statistical inhomogeneity and internal-wave spectra. However, we have not treated the effect of horizontal currents on sound transmission. Internal waves have associated horizontal currents whose effects (acting through the Doppler effect) are smaller than the direct sound-speed fluctuations, though not completely negligible (see the v_x/C column in Table 3.1).

CHAPTER 6

TRANSMISSION THROUGH A HOMOGENEOUS, ISOTROPIC MEDIUM

We begin this chapter by defining the quantitative nature of the sound-speed fluctuations in a manner that has been found to be most suitable for propagation theories; this involves correlation functions and spectral functions (§ 6.1). We describe the general case of anisotropic, inhomogeneous fluctuations before specializing to isotropy and homogeneity. We then describe the important parameters for wave propagation in homogeneous, isotropic random media, and discuss the different parameter regimes (§ 6.2). The geometrical-optics regime is covered in § 6.3 including descriptions of the important coherence functions used for describing the fluctuation effects. In § 6.4 some of the known results for regimes other than that of geometrical optics are discussed.

6.1. Correlation functions and spectral functions

The oceanic sound speed as a function of space and time can be expressed as:

$$C(\mathbf{x}, t) = C_0[1 + U(\mathbf{x}, t)] = C_0[1 + U_0(z) + \mu(\mathbf{x}, t)]. \quad (6.1.1)$$

The quantity C_0 is a constant, U_0 is a function only of vertical position, and μ is a function of space and time. The separation into these three terms is arbitrary, but the following discussion should make it clear why it is physically useful.

The sound speed in the ocean, if averaged over a long time, is principally a function of depth. Thus $C_0[1 + U_0(z)]$ can be considered as the local equilibrium sound channel (cf. Chapter 1) with C_0 taken for convenience as the sound speed at the sound axis.

We shall mean by μ, then, the deviation of the sound speed from its equilibrium value. By this definition μ does not include the effect of the deterministic sound channel, but only deviations (e.g. due to internal waves) from it. For later convenience we define:

$$\sigma = \text{acoustic (angular) frequency,} \quad \lambda = \text{wavelength}, \qquad (6.1.2)$$

$$q_0 = \sigma/C_0 = \text{acoustic wavenumber (to zeroth order)}, \qquad (6.1.3)$$

$$q(\mathbf{x}) \equiv (\sigma/C_0)[1 + U_0(z)]^{-1}$$

$$= \text{acoustic wavenumber averaged over time}, \qquad (6.1.4)$$

$$V(\mathbf{x}) = 2q^2\mu(\mathbf{x}, t), \qquad (6.1.5)$$

where the explicit dependence of V on time has been suppressed because it does not enter directly into the equations for sound propagation. (The quantity $V(\mathbf{x})$ varies so slowly with time that the sound-speed fluctuations may be considered as frozen during the passage of a sound pulse.) The quantity $V(\mathbf{x})$ is convenient only in Chapter 10, elsewhere $\mu(\mathbf{x}, t)$ will be used.

Correlation functions

By its definition

$$\langle\mu\rangle = 0, \qquad (6.1.6)$$

where the average is over a statistical ensemble, or over a long time. As discussed in Chapter 3, ocean fluctuations (such as internal waves) can be characterized by a statistical spectrum for μ or by a correlation function defined by

$$\rho[\mathbf{x}-\mathbf{x}', t-t', \tfrac{1}{2}(\mathbf{x}+\mathbf{x}')] = \langle\mu(\mathbf{x}, t)\mu(\mathbf{x}', t')\rangle. \qquad (6.1.7)$$

In fact, the dependence on the average position enters only through the vertical position

$$\bar{z} = \tfrac{1}{2}(z + z'), \qquad (6.1.8)$$

so that $\rho = \rho(x-x', y-y', z-z', t-t', \bar{z}) = \rho(\Delta\mathbf{x}, t-t', \bar{z})$ where $\Delta\mathbf{x} = (x-x', y-y', z-z')$.

Spectral functions

It will be useful to define the Fourier transform of ρ as F, in terms of a wavenumber, $\mathbf{k}$, conjugate to $\Delta\mathbf{x}$;

$$F(\mathbf{k}, t-t', \bar{z}) = (2\pi)^{-3} \int d^3(\Delta\mathbf{x}) e^{-i\mathbf{k}\cdot\Delta\mathbf{x}} \rho(\Delta\mathbf{x}, t-t', \bar{z}). \qquad (6.1.9)$$

The function F is called the spectral function. In all cases to be considered the time dependence of F (or ρ) may be eliminated because it is already implicit in the dependence on **k** (or Δ**x**), either through the frozen fluctuation hypothesis (Taylor's hypothesis) or through a dispersion relation. Hence we may write $\rho(\Delta\mathbf{x}, \bar{z})$ or $F(\mathbf{k}, \bar{z})$. The dependence on k_V may be eliminated in favor of a dependence on time; or (in Fourier-transform space) on frequency, ω, or (by considering discrete vertical modes) on mode number, j. Hence the various spectral functions are related by

$$d^3\mathbf{k}F(\mathbf{k}, \bar{z}) = d^2\mathbf{k}_H d\omega F(\mathbf{k}_H, \omega, \bar{z}) = d^2\mathbf{k}_H dj F(\mathbf{k}_H, j, \bar{z})$$
$$= dk\ d\omega(d\theta/4\pi)F(k, \omega, \bar{z}) = dk(d\theta/4\pi)\,dj F(k, j, \bar{z})$$

$$(6.1.10)$$

where $k = |\mathbf{k}_H|$ is the magnitude of the horizontal wavenumber, θ measures the direction of $\mathbf{k}_H$, and $dj = 1$. Note that we consider only spectra that are isotropic in the horizontal plane. If the space and time variables in ρ were independent, we would have had to do another Fourier transform to obtain $F(\mathbf{k}, \omega)$.

Correlation lengths

In general the definition of correlation lengths is difficult (and arbitrary) because the value obtained may be crucially dependent on the method of definition. For example, if an average of some power of k over the spectrum is used, then $\langle k^2\rangle^{-\frac{1}{2}}$ may be completely different from $\langle k^4\rangle^{-\frac{1}{4}}$; in fact one may be undefined or infinite while the other is not. We prefer to define correlation lengths and times in terms of the behavior of ρ at small separations in space and time, since that is what is usually experimentally measured. That is, for small $\Delta\mathbf{x} = (\Delta x, \Delta y, \Delta z)$ and $\Delta t = t - t'$ we may expand $\rho(\Delta\mathbf{x}, \Delta t, \bar{z}) = \rho(\Delta x, \Delta y, \Delta z, \Delta t; \bar{z})$ as follows;

$$\rho(0, 0, \Delta z, 0; \bar{z}) = \langle\mu^2(\bar{z})\rangle(1 - |\Delta z/L_V|^{p-1}), \qquad (6.1.11)$$

$$\rho(\Delta x, 0, 0, 0; \bar{z}) = \langle\mu^2(\bar{z})\rangle(1 - |\Delta x/L_H|^{q-1}), \qquad (6.1.12)$$

$$\rho(0, 0, 0, \Delta t; \bar{z}) = \langle\mu^2(\bar{z})\rangle(1 - |\Delta t/\tau|^{w-1}). \qquad (6.1.13)$$

The quantities L_H, L_V, and τ are functions of vertical position $\bar{z}$. The power-law exponents p, q, and w may be different and also may be functions of $\bar{z}$, although it should be emphasized again that the dependence on $\bar{z}$ is much weaker than the dependences on $\Delta\mathbf{x}$.

Homogeneous, isotropic fluctuations

Most of the previous literature is based on homogeneous isotropic fluctuations, particularly turbulence. In this case we may drop the dependence on $\bar{z}$, and write

$$L = L_H = L_V, \quad q = p. \tag{6.1.14}$$

The correlation function ρ is a function only of the magnitude of the separation $r = |\Delta \mathbf{x}|$, and we may write a one-dimensional spectral function as though we made measurements of the correlation function along some straight line in space:

$$F_1(k) = (2\pi)^{-1} \int_{-\infty}^{\infty} dr \, e^{-ikr} \rho(r). \tag{6.1.15}$$

The three-dimensional spectral function is then given by

$$F(\mathbf{k}) = -(2\pi k)^{-1} \partial_k F_1(k). \tag{6.1.16}$$

Turbulence

Tatarskii (1971) shows that for turbulence in the inertial subrange

$$p = \tfrac{5}{3}, \quad C_n^2 = 2\langle \mu^2 \rangle L^{-\frac{2}{3}},$$

$$F(\mathbf{k}) = 0.033 \, C_n^2 k^{-\frac{11}{3}}, \quad F_1(k) = 0.124 \, C_n^2 k^{-\frac{5}{3}}, \tag{6.1.17}$$

in the region $L_0^{-1} < k < 2\pi/\lambda_0$ where λ_0 and L_0 are the inner and outer scales of turbulence. Note that the form of the exponent in (6.1.11) to (6.1.13) was dictated by the desire that $F_1(k) \sim k^{-p}$.

6.2. Parameters and regimes; $\Lambda - \Phi$ space

The character of the fluctuations in a wave field that has passed through isotropic inhomogeneities (in an otherwise constant index of refraction) is controlled by two parameters representing the *strength* and *size* (spatial extent) of the inhomogeneities. (Hereafter, to maintain our emphasis on ocean acoustics, we refer to sound-speed fluctuations; this corresponds directly to index-of-refraction fluctuations.)

To characterize *size* we define Λ such that

$$\Lambda \propto (R_F/L)^2. \tag{6.2.1}$$

The quantity R_F is the radius of the first Fresnel zone for a source and receiver separated by range R, and L is the correlation length

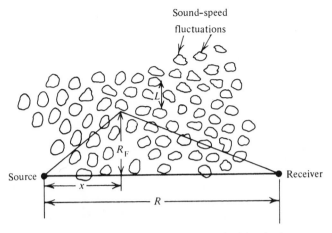

Fig. 6.1. Definition of the Fresnel-zone radius (R_F) in the homogeneous isotropic case.

(see Fig. 6.1). A small value of Λ corresponds to small effects from diffraction (so that geometrical optics applies), while in the region of large Λ diffraction is important. Note that Λ does *not* characterize the spatial extent of the sound-speed fluctuations (L does that) but rather the diffraction effect caused by the spatial extent. That is the reason for the inverse relation between Λ and L.

To define Λ more precisely we note that the radius of a Fresnel zone is a function of the range x at which a scattering takes place. Hence we should average $[R_F(x)/L]^2$ over the straight line connecting source and receiver (see Fig. 6.1). A simple analysis that makes use of the fact that $R_F \ll R$ for all cases of interest yields

$$R_F^2(x) = \lambda x(R-x)/R. \qquad (6.2.2)$$

We still have an arbitrary overall constant to insert for the normalization of Λ. It is convenient to define Λ with a factor of 2π inserted (see Chapter 8). Thus:

$$\Lambda = R^{-1} \int_0^R (2\pi)^{-1}[R_F(x)/L]^2 \, dx = R/6L^2 q_0. \qquad (6.2.3)$$

The quantity Λ is closely related to the *wave parameter* described by Tatarskii (1971). We refer to Λ as the *diffraction parameter*.

To characterize *strength* we integrate the sound-speed fluctuations from source to receiver along a straight line and take an average, defining Φ such that

$$\Phi^2 \equiv \left\langle \left(q_0 \int_0^R \mu \, dx \right)^2 \right\rangle = q_0^2 \int_0^R \int_0^R dx \, dx' \rho(|x - x'|). \quad (6.2.4)$$

For the usual case of $R \gg L$ we may write

$$\Phi^2 = q_0^2 \int_0^R dx \left[\int_{-\infty}^{\infty} \rho(u) \, du \right], \quad (6.2.5)$$

which can be approximately evaluated using (6.1.12):

$$\Phi^2 \approx q_0^2 \langle \mu^2 \rangle R L_P,$$
$$L_P = \langle \mu^2 \rangle^{-1} \int_{-\infty}^{\infty} \rho(u) \, du \approx 0.4L. \quad (6.2.6)$$

Note that the integral scale length, L_P, enters in the expression for Φ^2, and it is not necessarily equal to the differential scale length defined by (6.1.12).

The quantity Φ is the rms variation in phase of a signal at the receiver under the geometrical optics approximation.

Other workers have defined the Λ and Φ parameters (or their equivalents) and their regimes (see particularly de Wolf, 1975). It is important to note that knowledge of the sound-speed fluctuations in the form of $\rho(r)$ is all that is required to determine the values of Φ and Λ; no wave-propagation measurements need to be made. We refer to Φ as the *strength parameter*.

Fig. 6.2 shows a graphical representation of Λ–Φ space with several important sound-transmission regimes outlined. The heavy lines form the most important boundary on the diagram – the boundary between unsaturated (weak) scattering and saturated (strong) scattering. The boundary between the two is characterized by sound intensity fluctuations reaching unity (Tatarskii, 1971).

Again following Tatarskii (1971), but retaining an arbitrary power-law exponent, p, it can be shown that the condition for the boundary for small Λ is

$$\Phi \Lambda^\alpha = 1 \quad (6.2.7)$$

where $\alpha = p/4$ if $p < 4$ and $\alpha = 1$ if $p > 4$. Thus the boundary is a straight line for log–log scaling. Table 6.1 shows (6.2.7) evaluated for isotropic fluctuations, for several interesting power-law

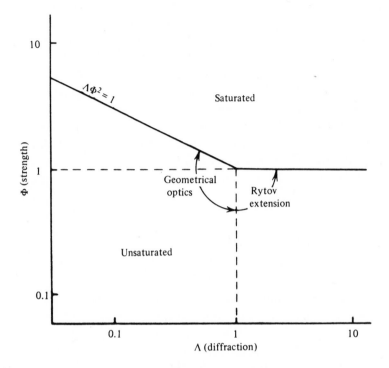

Fig. 6.2. Regimes of Λ–Φ space.

Table 6.1. *The requirement for weak scattering in the geometrical-optics region ($\Lambda <$ 1) in the case of isotropic fluctuations with a spectrum $F_1(k) \sim k^{-p}$ and a correlation length L; R is the range and λ is the acoustic wavelength.*

p	$\Phi \Lambda^{p/4} < 1$
$\frac{5}{3}$	$\mu_{\mathrm{rms}} < 6^{\frac{5}{12}} q^{-\frac{7}{12}} L^{\frac{1}{3}} R^{-\frac{11}{12}}$
2	$\mu_{\mathrm{rms}} < (6L/q_0)^{\frac{1}{2}}/R$
$\geqslant 4$	$\mu_{\mathrm{rms}} < 6(L/R)^{\frac{3}{2}}$
(includes constant scale size case)	

exponents. In the diffraction region ($\Lambda > 1$) the boundary is just the line $\Phi = 1$; that is, when the rms phase fluctuations approach unity in this region, so do the intensity fluctuations.

Having established the boundaries, the task of any complete sound-transmission theory is to describe the type of signal characteristics in each regime, but even more importantly to calculate quantitatively the parameters (such as coherence lengths and times) of the signal from a knowledge of ocean structure. In the following sections we discuss present knowledge for the isotropic turbulence case, in preparation for our generalization to a realistic ocean in Chapters 7 and 8.

6.3. Geometrical optics

In the unsaturated regime of Fig. 6.2, a subregion exists for $\Lambda < 1$. This subregion has been analyzed extensively (see for example, Tatarskii, 1971) and is called the geometrical-optics regime. The bordering, non-geometrical-optics regions are either dominated by diffraction ($\Lambda > 1$) or by saturation phenomena that occur for intensity fluctuations approaching unity.

In the geometrical-optics region the wave-function may be written following (5.4.3);

$$\psi(\mathbf{x}_1, \mathbf{x}_2) = K(\mathbf{x}_1, \mathbf{x}_2) \exp\left[iq_0 S(\mathbf{x}_1, \mathbf{x}_2)\right], \qquad (6.3.1)$$

which we may also express as:

$$\psi(\mathbf{x}_1, \mathbf{x}_2) = \psi_0(\mathbf{x}_1, \mathbf{x}_2) \exp\left[iq_0 \int_0^R \mu(x') \, dx'\right], \qquad (6.3.2)$$

where $\psi_0(\mathbf{x}_1, \mathbf{x}_2)$ is the wavefunction obtained with no sound-speed fluctuations and x' is a coordinate along a straight line from the source at $\mathbf{x}_1$ to the receiver at $\mathbf{x}_2$.

The wavefunction ψ has a phase that is different from its unperturbed phase by an amount

$$\phi(\mathbf{x}) = q_0 \int_0^R \mu(x') \, dx', \qquad (6.3.3)$$

where $\mathbf{x}$ is the position of the receiver, and the dependence on the source position has been suppressed. We see immediately from

(6.2.4) that the strength parameter Φ is, in the geometrical-optics regime, the rms phase fluctuation of the received signal:

$$\langle[\phi(\mathbf{x})]^2\rangle = \left\langle\left[q_0\int_0^R \mu(\mathbf{x}')\,\mathrm{d}x'\right]^2\right\rangle \equiv \Phi^2. \qquad (6.3.4)$$

The expectation value of the received signal is

$$\langle\psi\rangle = \psi_0\langle\exp[i\phi(\mathbf{x})]\rangle. \qquad (6.3.5)$$

The function $\mu(\mathbf{x})$ is a Gaussian random variable; therefore so is $\phi(\mathbf{x})$, and the rules of Gaussian statistics tell us that

$$\langle\psi\rangle = \psi_0\exp\{-\tfrac{1}{2}\langle[\phi(\mathbf{x})]^2\rangle\} = \psi_0\exp(-\tfrac{1}{2}\Phi^2). \qquad (6.3.6)$$

Using the approximation of (6.2.6) we can thus write

$$\langle\psi\rangle = \psi_0\exp(-R/L_E), \quad L_E^{-1}\approx\tfrac{1}{2}q_0^2\langle\mu^2\rangle L, \qquad (6.3.7)$$

where L_E is the *extinction length*.

Coherences and the phase-structure function

It is well known (Tatarskii, 1971) that the fluctuations in the acoustic path difference between two receiver locations are very important in the consideration of coherence functions for received signals. Fig. 6.3 shows a source and two receivers separated vertically by a distance Δz. The phase-structure function is defined as

$$D(\Delta z) \equiv \left\langle\left[q_0\int_0^{\mathbf{x}_1}\mu(\mathbf{x})\,\mathrm{d}x - q_0\int_0^{\mathbf{x}_2}\mu(\mathbf{x}')\,\mathrm{d}x'\right]^2\right\rangle. \qquad (6.3.8)$$

In the homogeneous isotropic case the above expression may be evaluated in terms of the spectral function of the sound-speed fluctuations. First we note that $D(\Delta z)$ may be expressed as an integral over the straight line between source and receiver. We define a new function $d(\Delta z)$ which can be identified as the local contribution to $D(\Delta z)$ at range x:

$$D(\Delta z) = \int_0^R d(\Delta z')\,\mathrm{d}x, \quad \Delta z' = x\,\Delta z/R. \qquad (6.3.9)$$

Thus the integral represents an average of a function of the separation $\Delta z'$ that begins with zero and ends with Δz. One can then show that

$$d(\Delta z') = 8\pi^2 q_0^2\int_0^\infty [1 - J_0(k\,\Delta z')]F(k)k\,\mathrm{d}k \qquad (6.3.10)$$

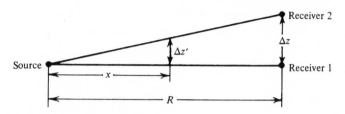

Fig. 6.3. Rays from a source to two receivers separated by a vertical distance Δz.

where J_0 is the Bessel function of zero order and $F(k)$ is the three-dimensional spectral function of the sound-speed fluctuations.

The calculation of $D(\Delta z)$ depends only on the sound-speed fluctuations and not on sound propagation; its importance for sound propagation is illustrated by considering two experimentally interesting observables in the geometrical-optics regime.

The first example is the phase difference between the signals at the first and second receivers in Fig. 6.3. We may write immediately from (6.3.3) and (6.3.8):

$$\langle [\phi(\mathbf{x}_1) - \phi(\mathbf{x}_2)]^2 \rangle = D(\Delta z). \tag{6.3.11}$$

Hence the phase difference between two hydrophones, separated by a distance Δz transverse to the line of sound propagation, fluctuates with a variance given by $D(\Delta z)$.

The second example is the covariance between the wavefunctions at $\mathbf{x}_1$ and $\mathbf{x}_2$:

$$\langle \psi^*(\mathbf{x}_1)\psi(\mathbf{x}_2) \rangle = \psi_0^*(\mathbf{x}_1)\psi_0(\mathbf{x}_2)\langle \exp\left[i\phi(\mathbf{x}_2) - i\phi(\mathbf{x}_1)\right] \rangle.$$

$$\tag{6.3.12}$$

The rules of Gaussian statistics immediately yield:

$$\langle \psi^*(\mathbf{x}_1)\psi(\mathbf{x}_2) \rangle = \psi_0^*(\mathbf{x}_1)\psi_0(\mathbf{x}_2) \exp\left[-\tfrac{1}{2}D(\Delta z)\right]. \tag{6.3.13}$$

Thus the experimentally measurable sound-transmission characteristics may be calculated from the sound-speed fluctuation spectra through the phase-structure function $D(\Delta z)$.

We may calculate $D(\Delta z)$ for the case of isotropic turbulence, using the spectrum (6.1.17) with a cutoff at high wavenumber. We

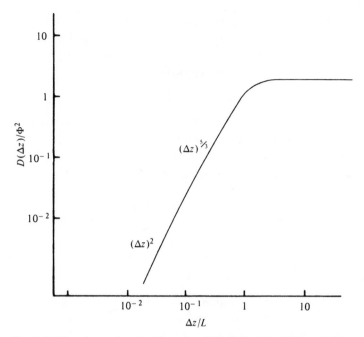

Fig. 6.4. The phase-structure function $D(\Delta z)$ for isotropic turbulence.

begin with the result for d

$$d(\Delta z) \approx \begin{cases} 3.28\langle\mu^2\rangle q_0^2 L^{\frac{4}{3}}\lambda_0^{-\frac{1}{3}}(\Delta z/L)^2, & \Delta z \ll \lambda_0 \\ 5.84\langle\mu^2\rangle q_0^2 L(\Delta z/L)^{\frac{5}{3}}, & \lambda_0 \ll \Delta z \ll L_0 \\ 2\langle\mu^2\rangle q_0^2 L, & \Delta z \gg L_0. \end{cases} \quad (6.3.14)$$

The last results follow from the independence of fluctuations at two widely separated points. $D(\Delta z)$ is then obtained by integration (Fig. 6.4);

$$D(\Delta z) = \begin{cases} \frac{1}{3}Rd(\Delta z), & \Delta z \ll \lambda_0 \\ \frac{3}{8}Rd(\Delta z), & \lambda_0 \ll \Delta z \ll L_0 \\ Rd(\Delta z), & \Delta z \gg L_0. \end{cases} \quad (6.3.15)$$

Equation (6.3.15) can also be expressed as

$$D(\Delta z) = \begin{cases} 1.1\Phi^2(L/\lambda_0)^{\frac{1}{3}}(\Delta z/L)^2, & \Delta z \ll \lambda_0 \\ 2.2\Phi^2(\Delta z/L)^{\frac{5}{3}}, & \lambda_0 \ll \Delta z \ll L_0 \\ 2\Phi^2, & \Delta z \gg L_0. \end{cases} \quad (6.3.16)$$

We have considered only a spatial separation between the points x_1 and x_2, but we could have equally well considered a time separation. The result for turbulence, which depends on the Taylor hypothesis of convected, frozen turbulence, is

$$d(\Delta t) = d(\Delta z) \text{ for } \Delta z = v\,\Delta t \tag{6.3.17}$$

where Δt is the time separation and v is the convection velocity.

Frequency separation is of interest when a broad-band signal is transmitted from a source to a given receiver. The phase-structure function for a frequency separation $\Delta \sigma = \sigma - \sigma'$ is

$$D(\Delta \sigma) = \langle [(\sigma/C_0) \int_0^{x_1} \mu\,dx - (\sigma'/C_0) \int_0^{x_1} \mu\,dx]^2 \rangle. \tag{6.3.18}$$

Using (6.3.4)

$$D(\Delta \sigma) = (\Delta \sigma/\sigma)^2 \Phi^2. \tag{6.3.19}$$

We have treated correlations of complex amplitude and phase for separations in space, time, and frequency separately. For small displacements in position, time, and frequency

$$D(\Delta z, \Delta t, \Delta \sigma) = D(\Delta z) + D(\Delta t) + D(\Delta \sigma), \tag{6.3.20}$$

and in analogy to (6.3.13), the generalized covariance of the complex amplitude at two separated points is given by

$$\langle \psi^*(\Delta z, \Delta t, \Delta \sigma)\psi(0, 0, 0) \rangle = |\psi_0|^2 \exp\left[-\tfrac{1}{2}D(\Delta z, \Delta t, \Delta \sigma)\right]. \tag{6.3.21}$$

These results from geometrical optics in a homogeneous, isotropic medium will be generalized to the ocean in Chapter 8. In fact, with proper definitions of Φ, Λ, and D, many of the equations will be unchanged and will even be applicable beyond the regime of geometrical optics.

Intensity fluctuations

Intensity fluctuations and correlations are much more sensitive to the details of the sound-speed fluctuation spectrum than the phase or complex-amplitude fluctuations. Power-law spectra cause divergent integrals that must be treated carefully for each special case; for example in the turbulence case the amplitude fluctuations depend crucially on the sound-speed spectrum in the region of the inner scale of turbulence. Further discussion of intensity fluctuations is deferred to the ocean case in Chapter 8 and Part IV.

6.4. Other parameter regimes

The basic equation of geometrical optics is (6.3.1) whose range of validity is known (Fig. 6.2). However, it is also known that many of the formulas originally derived from geometrical optics have a range of validity for homogeneous, isotropic media that far exceeds that of (6.3.1). The section of the unsaturated regime where diffraction is important has been investigated thoroughly. The most valuable method has been the Rytov method that applies to the entire unsaturated region, including geometrical optics. On the other hand the saturated region has proven more intractable, and only a few results are previously known or calculable.

The most important formula in § 6.3 for analyzing signal statistics is (6.3.21). This equation has been shown to be valid throughout the unsaturated region, and has also been shown to be valid in the saturated region using the method of moments (Tatarskii, 1971, and Beran, 1970). We show its general validity for the ocean case with $\Delta\sigma = 0$ in Part IV. The result for $\Delta\sigma \neq 0$ must be modified in the saturated region (see Chapter 8).

On the other hand, the validity of (6.3.11) does not extend beyond the geometrical-optics regime. In the unsaturated region in the limit of large Λ one can show that

$$\langle[\phi(\mathbf{x}_1) - \phi(\mathbf{x}_2)]^2\rangle = \tfrac{1}{2}D(\Delta z) \tag{6.4.1}$$

and in the saturated region the statistics of phase bear no relation to the phase-structure function, because of many-ray interference.

CHAPTER 7

THE OCEAN MEDIUM

Four important differences distinguish the ocean medium from a medium with homogeneous isotropic turbulence: anisotropy, statistical inhomogeneity, the deterministic sound channel, and a different spectral function.

The ocean is *anisotropic*. The internal-wave spectrum is such that the correlation lengths of sound-speed fluctuations are very different in the horizontal and vertical; in fact $L_H \gg L_V$. The phase-structure function therefore depends on the orientation of the ray from source to receiver.

The ocean is *statistically inhomogeneous*. The value of $\langle \mu^2 \rangle$ is a function of depth, being much larger near the surface than at great depths (see (1.6.10)). Hence evaluations of averages along a ray must account for the fact that contributions from excursions near the surface will be much larger than (and may dominate over) contributions from excursions to great depths. Also the frequency of internal waves contributing to the spectrum at a given depth z is restricted to be less than $n(z)$. Through the dispersion relation this acts as a restriction on the relation between horizontal and vertical wavenumbers and hence causes correlation lengths to depend on depth.

The ocean has a *sound channel*. Thus the unperturbed rays are curved (see Chapter 4) making averages along rays even more complicated. In addition, the concept of a Fresnel zone, so crucial to the definition of the sound-speed-fluctuation size parameter, requires considerable amplification. Finally, the curved rays lead to the phenomenon of *deterministic multipath*, which complicates received signals.

The ocean variability is characterized by a *spectrum* in space and time that does not correspond to turbulence. Whenever a specific spectrum is needed, we shall for concreteness assume a *spectrum of internal waves* in the region of scale sizes (≥ 1 m) of interest to long-range sound propagation. The vertical and horizontal spectra are both characterized by a power-law exponent $p = 2$ in contrast to $p = \frac{5}{3}$ for turbulence. We must also emphasize that the internal-wave spectrum yields an intrinsic time dependence. Turbulence is usually considered as frozen, with a time dependence coming only if the entire turbulent field is being convected past the source and receiver (Taylor's hypothesis).

These differences from a homogeneous, isotropic turbulent medium require a careful re-analysis of several quantities already seen to be important in the last chapter: Fresnel zones (§ 7.1), the size and strength parameters Λ and Φ (§ 7.2), and the phase-structure function (§ 7.3). Specific formulas using the internal-wave spectrum are given in §§ 7.4 and 7.5.

7.1. Fresnel zones and ray tubes

As shown in Chapter 6, the size of a Fresnel zone for a particular source–receiver geometry is crucial to the understanding of fluctuations. The concept of a Fresnel zone in ocean sound transmission is complicated by the presence of the *sound channel*.

Consider a ray in the deterministic sound channel with no fluctuations from internal waves or other disturbances (Fig. 7.1). Following (5.4.4), the acoustic path length along the ray is

$$S_0(0, \mathbf{x_R}) = \int_0^{\mathbf{x_R}} \{\tfrac{1}{2}(y_R/R)^2 + \tfrac{1}{2}[\partial_x z_{\text{ray}}(x)]^2 - U_0\} \, dx, \quad (7.1.1)$$

where the subscript zero indicates the path length undisturbed by fluctuations.

Suppose we choose an intermediate point $\mathbf{x}$ lying on the ray somewhere between the source and the receiver, and we displace it to a point $\mathbf{x} + \Delta\mathbf{x}$. We may then look at two other deterministic rays; one joining the source to $\mathbf{x} + \Delta\mathbf{x}$ and the other joining $\mathbf{x} + \Delta\mathbf{x}$ to the receiver. This geometry describes the propagation of a signal from the source to $\mathbf{x} + \Delta\mathbf{x}$, at which point a scattering due to a fluctuation occurs, and the subsequent propagation to the receiver. The total acoustic path length for these two rays is $S_0(0, \mathbf{x} + \Delta\mathbf{x}) + S_0(\mathbf{x} + \Delta\mathbf{x}, \mathbf{x_R})$, and this evidently differs from the unscattered path

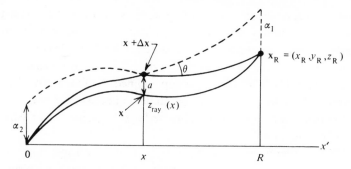

Fig. 7.1. Geometry for defining Fresnel zones and ray tubes in the presence of a deterministic sound channel.

length $S_0(\mathbf{0}, \mathbf{x_R})$. The derivation of the scattered path from the unperturbed ray is related to two equivalent geometrical concepts: Fresnel zones or ray tubes.

The Fresnel-zone size, R_F, is defined to be the value of Δx such that the scattered path length deviates from the unscattered one by half a wavelength. This size depends on the orientation of Δx; we choose R_F to be the value for a purely vertical displacement $\Delta \mathbf{x} = (0, 0, a)$. When a is small we have

$$S_0(\mathbf{0}, \mathbf{x} + \Delta \mathbf{x}) + S_0(\mathbf{x} + \Delta \mathbf{x}, \mathbf{x_R}) - S_0(\mathbf{0}, \mathbf{x_R})$$
$$= \tfrac{1}{2}a^2 \partial_{zz}[S_0(\mathbf{0}, \mathbf{x}) + S_0(\mathbf{x}, \mathbf{x_R})]. \qquad (7.1.2)$$

We define the *phase curvature*, $A(x)$, by

$$A(x) = \partial_{zz}[S_0(\mathbf{0}, \mathbf{x}) + S_0(\mathbf{x}, \mathbf{x_R})] \qquad (7.1.3)$$

so that $A(x)$ is the second derivative of the path length for a small vertical displacement at a range x. A measure of the rate at which the path length changes as one moves away from a ray is given by $A(x)$. By definition of R_F,

$$AR_F^2/2 = \lambda/2, \qquad (7.1.4)$$

so that

$$R_F^2(x) = \lambda A^{-1}, \quad R_F^2(x)/2\pi = (q_0 A)^{-1}. \qquad (7.1.5)$$

For a displacement $\Delta \mathbf{x}$ in an arbitrary direction we define a matrix

$$A_{ij}(x) = \partial_i \partial_j[S_0(\mathbf{0}, \mathbf{x}) + S_0(\mathbf{x}, \mathbf{x_R})], \qquad (7.1.6)$$

where i and j denote components of $\mathbf{x}$. This matrix has the form

$$A_{ij}(x) = \begin{pmatrix} A(x)\tan^2\theta(x) & 0 & -A(x)\tan\theta(x) \\ 0 & R/x(R-x) & 0 \\ -A(x)\tan\theta(x) & 0 & A(x) \end{pmatrix}$$

(7.1.7)

where $\theta(x)$ is the angle of the unperturbed ray with the horizontal at x.

The *phase curvature* can also be derived by using the concept of a *ray tube*, consisting of nearby rays surrounding our unperturbed ray. We begin by examining the differential equation satisfied by rays close to the original ray $z_{ray}(x)$. Let a nearby ray be $z(x) = z_{ray}(x) + \xi(x)$ where $\xi(x)$ is small. The linearized equation for ξ is

$$\partial_{xx}\xi(x) + U_0''[z_{ray}(x)]\xi(x) = 0. \qquad (7.1.8)$$

We define the two solutions of this second-order differential equation by the following boundary conditions: $\xi_1(x)$ is defined by $\xi_1(0) = 0$, $\xi_1(R) = 1$; whereas $\xi_2(x)$ is defined by $\xi_2(0) = 1$, $\xi_2(R) = 0$. It is clear from Fig. 7.1 that the displacement of the perturbed path from the unperturbed ray is given by $\alpha_1\xi_1(x')$ from $x' = 0$ to $x' = x$ and $\alpha_2\xi_2(x')$ from $x' = x$ to $x' = R$. Thus we also have $\alpha_1\xi_1(x) = \alpha_2\xi_2(x)$. It is then simple to show that the angle θ is given by

$$\theta(x) = \frac{\xi_2(x)\,\partial_x\xi_1(x) - \xi_1(x)\,\partial_x\xi_2(x)}{\xi_1(x)\xi_2(x)}a. \qquad (7.1.9)$$

In this picture, θ is the sum of the two ray-tube angles, one tube from the left and the other from the right, at position x. The important quantity that measures the geometry of the situation is of course the coefficient of the small displacement a, in other words $d\theta/da$. Use of (7.1.1) establishes that $d\theta/da \equiv A(x)$, so that

$$A(x) = \frac{\xi_2(x)\,\partial_x\xi_1(x) - \xi_1(x)\,\partial_x\xi_2(x)}{\xi_1(x)\xi_2(x)}. \qquad (7.1.10)$$

Thus $A(x)$ is the rate of change of the sum of ray-tube angles as a function of the displacement away from the unperturbed ray.

Approximations to $A(x)$

There are a few regimes where A may be obtained without specific assumptions regarding the form of the sound channel.

First, for very deep rays, the sound channel varies nearly linearly with depth and the rays are nearly arcs of circles. For this case, to first order in the radius of curvature, the quantity A has the same value as in a homogeneous ocean:

$$A^{-1} = x(R-x)/R. \tag{7.1.11}$$

Second, for near-axis rays, the sound channel is nearly a parabolic function of depth, so that the rays are nearly sinusoidal. Then we find

$$A^{-1} = \left| \frac{\sin Kx \sin K(R-x)}{K \sin KR} \right|, \tag{7.1.12}$$

where $2\pi/K$ is the wavelength of the sinusoidal ray, which we label as R_D, the range of a double-axis loop. Note that this approximation to A^{-1} becomes infinite when the receiver is located at axis crossings of the ray. These points are caustics for sinusoidal rays. Caustics are, in general, points at which A^{-1} diverges, that is, where $A = 0$. When this occurs, the matrix A_{ij} takes the simple form:

$$A_{ij} = \begin{pmatrix} 0 & 0 & 0 \\ 0 & R/x(R-x) & 0 \\ 0 & 0 & 0 \end{pmatrix}, \tag{7.1.13}$$

with the result that the second-order term in the transverse derivative of the optical path no longer dominates. To calculate fluctuations correctly in this region, we would have to keep third-order derivatives of $S(0, y) + S(y, y_1)$. Presumably, there is in fact some unusual structure at these points, but our theories, at the level to which we have carried them, cannot correctly describe this structure. Even more significant however is that for near-axis rays a Fresnel-zone radius does *not* gradually increase with range, even on the average, but rather oscillates around values typical for ranges of only $R_D/2 \sim 20$ km!

A third regime in which we can obtain A^{-1} without knowledge of the details of the sound channel is that of very long ranges, in which the rays contain a large number of loops. In this case, Munk and Zachariasen (1976) have shown that

$$A^{-1} = (\tan^2 \theta/\delta)[x(R-x)/R] \tag{7.1.14}$$

where δ is given by (4.1.11).

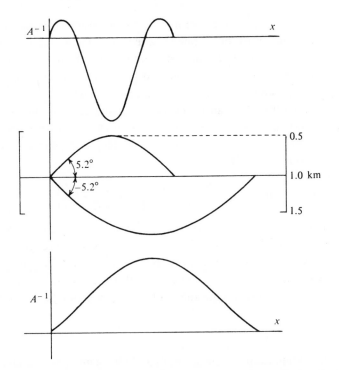

Fig. 7.2. A^{-1} for 5.2° upward loop (top) and downward loop (bottom), with the ±5.2° ray itself shown in the center, all plotted as functions of horizontal distance x.

This last formula illustrates that, just as in the case of near-axis rays, the radius of a Fresnel zone (7.1.5) may vary considerably from its value in a homogeneous ocean (7.1.11). For the canonical profile (1.6.6) the value of $|\delta|$ is ~0.07 whereas $\tan^2 \theta$ varies between zero at turning points and ~0.04 at the sound-axis depth. Thus the Fresnel-zone radius is always smaller in the ocean than in a homogeneous medium.

In the general case one can, of course, numerically compute A for any given source, receiver, and ray configuration. Fig. 7.2 shows two examples of A^{-1} as a function of x using the canonical profile.

The quantity B_{ij}

The 'phase curvature' A was defined for a point source and a point receiver. However, in certain cases of interest the receiver may not be a point, but rather may be a vertical array. In that case a new quantity must be defined, called B_{ij}, which is the second derivative matrix of the acoustic path length from a source to receiver located at a fixed range and seeing a fixed vertical angle, rather than one located at a fixed range and height.

The quantity B^{-1} can be evaluated analytically for linear and quadratic sound channels (with circular and sinusoidal rays) which approximate the real sound channel for deep and near-axial rays respectively. We find

$$B^{-1} = x, \quad B^{-1} = \left| \frac{\sin Kx \cos K(R-x)}{K \cos KR} \right| \qquad (7.1.15)$$

for deep and near-axial rays respectively, where $2\pi/K$ is the wavelength of the sinusoidal rays, so that π/K is the range of one loop. Thus, for near-axial rays, B^{-1} becomes infinite when the receiver is located at the turning point of the rays. It is here that all rays are parallel; this is the analogue of a caustic for a beam-former receiver.

7.2. Definitions of the strength and diffraction parameters, Φ and Λ

The *strength* parameter, Φ, for a homogeneous, isotropic medium was shown in Chapter 6 to be the rms fluctuation in an integral of the sound-speed fluctuations along the straight line connecting source and receiver. This integral is the phase fluctuation of the received signal in the geometrical-optics regime. The natural generalization to the ocean medium is an integral along the unperturbed ray from source to receiver (see Fig. 7.3):

$$\Phi^2 = \left\langle \left[q_0 \int_{\text{ray}} \mu(s) \, ds \right]^2 \right\rangle = q_0^2 \int_{\text{ray}} ds \int_{\text{ray}} ds' \rho(s, s'). \qquad (7.2.1)$$

The integrals are understood to be along the unperturbed ray with $\rho(s, s')$ being the correlation function of two points (Fig. 7.3). Changing variables to $x = \frac{1}{2}(x_1 + x_2)$ and $u = s - s'$, where u is understood to move tangentially to the ray at range x, we may write

$$\Phi^2 = q_0^2 \int_0^R dx \left\{ \int_{-\infty}^{\infty} du \, \rho \left[u, \theta, z_{\text{ray}}(x) \right] \right\}. \qquad (7.2.2)$$

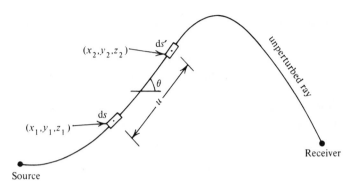

Fig. 7.3. Geometry for defining the strength parameter Φ in the presence of a deterministic sound channel.

The arguments of ρ are to indicate that the size and rate of decrease of the correlation function with distance u depends on the orientation of the line of integration (at angle θ) and on the depth of the ray at x. To make (7.2.2) more transparent we define a correlation length, $L_P(\theta, z)$, (P refers to the correlation length *parallel* to the ray) of the sound-speed fluctuations along a line oriented at angle θ and at depth z:

$$L_P(\theta, z) = \langle \mu^2(z) \rangle^{-1} \int_{-\infty}^{\infty} \rho(u, \theta, z) \, du. \qquad (7.2.3)$$

Then the strength parameter may be written:

$$\Phi^2 = q_0^2 \int_0^R dx \langle \mu^2(z_{\text{ray}}) \rangle L_P(\theta, z_{\text{ray}}). \qquad (7.2.4)$$

Equation (7.2.4) may be interpreted as follows. Each increment in range (dx) contributes to the strength of the signal fluctuations an amount proportional to the integrand in (7.2.4). This integrand is proportional to the variance of μ at that point on the ray, and also to the correlation length of the sound-speed fluctuations along the direction of the ray at that point. Thus (7.2.4) is affected by all four special properties of the ocean medium: anisotropy enters because L_P is a function of the orientation of the ray; statistical inhomogeneity enters because $\langle \mu^2 \rangle$ and L_P depend on depth; the sound channel enters because the ray is not a straight line; and the different spectrum controls the explicit evaluation of $\langle \mu^2 \rangle$ and L_P.

The *diffraction* parameter, Λ, for a homogeneous, isotropic medium, was defined in Chapter 6 as an average of $(R_F/L)^2/2\pi$ where R_F is a Fresnel-zone radius and L is the correlation length of ρ. Now, however, any average over a ray must use the weighting of the strength contribution of each point along the ray. In addition, the Fresnel-zone radius is now not a simple function of range, but involves the phase curvature. Hence we generalize Λ as:

$$\Lambda = \Phi^{-2}q_0^2 \int_0^R dx \langle \mu^2(z_{\text{ray}}) \rangle L_P(\theta, z_{\text{ray}}) |q_0 A L_V^2|^{-1}, \quad (7.2.5)$$

which is the proper average of $(q_0 A L_V^2)^{-1}$. The vertical correlation length, L_V, and the phase curvature, A, are both functions of range x. We have used L_V as the appropriate replacement for L because $A(x)$ (and hence a Fresnel zone) is defined for small *vertical* displacements from the unperturbed ray. A term in Λ associated with a Fresnel zone for *horizontal* displacements of the unperturbed ray has been neglected because $L_H \gg L_V$.

7.3. The phase-structure function, D

In analogy with the homogeneous, isotropic case (6.3.8) and (6.3.16), we define the phase-structure function as:

$$D(1, 2) = \left\langle \left[(\sigma_1/C_0) \int_1 \mu \, ds - (\sigma_2/C_0) \int_2 \mu \, ds' \right]^2 \right\rangle. \quad (7.3.1)$$

The numbers 1 and 2 refer to an integral along an unperturbed ray from the source to points in space $\mathbf{x}_1$ and $\mathbf{x}_2$, at times t_1 and t_2, and at frequencies σ_1 and σ_2 respectively. The geometry is indicated in Fig. 7.4. In the geometrical-optics regime, $D(1, 2)$ is the variance of the phase difference between points P_1 and P_2 caused by the sound-speed fluctuations.

The phase-structure function in the homogeneous, isotropic case is a function of three separation variables. In the ocean case it is a function of the unperturbed ray geometry as well as of the four separation variables: $\Delta z = z_1 - z_2$, $\Delta y = y_1 - y_2$, $\Delta t = t_1 - t_2$, and $\Delta\sigma = \sigma_1 - \sigma_2$. A complete formula for $D(1, 2)$ in terms of the

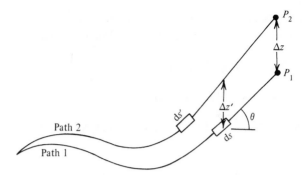

Fig. 7.4. Geometry for defining the phase-structure function D in the presence of a deterministic sound channel.

spectral function of sound-speed fluctuations is derived from (7.3.1):

$$D(\Delta z, \Delta y, \Delta t, \Delta \sigma) = (\Delta \sigma/\sigma)^2 \Phi^2 + \int_0^R d(\Delta z', \Delta y', \Delta t; z, \theta) \, dx,$$

$$(7.3.2)$$

where $\Delta z' = \Delta z \xi_1(x)$, $\Delta y' = (x/R) \Delta y$, and $d(\Delta z', \Delta y', \Delta t; z, \theta)$ is the contribution to the phase-structure function of the point at range x along the unperturbed ray. This contribution depends on the separation variables in space and time at that range x, and also on the depth and orientation (z and θ) of the ray at range x. We find

$$d(\Delta z', \Delta y', \Delta t; z, \theta) = 4\pi q_0^2 \int d\omega \int dk_V \int_0^\infty k_H \, dk_H \, k_y^{-1}$$

$$\cdot [1 - \cos(k_y \Delta y' + k_V \Delta z') \cos \omega \Delta t] F(\mathbf{k}, \omega, z), \quad (7.3.3)$$

where $k_y^2 = k_H^2 - k_V^2 \tan^2\theta$. The spectral function $F(\mathbf{k}, \omega, z)$ is written generally, although for internal waves only two of the three variables k_H, k_V and ω are independent.

We may expand the cosines for small displacements, but the evaluation of d will then depend on the power law exponent of the spectral function F. The result for the internal-wave model is given in the next section.

For a homogeneous, isotropic medium with $\Delta t = 0$, we set $\Delta y = 0$ and $\theta = 0$ without loss of generality, and integrate over ω:

$$d(\Delta z') = 4\pi q_0^2 \int dk_V \int dk_H (1 - \cos k_V \Delta z')F(k), \quad (7.3.4)$$

where $k^2 = k_V^2 + k_H^2$. We may then write

$$d(\Delta z') = 4\pi q_0^2 \int_0^{2\pi} d\theta \int_0^\infty k \, dk \, [1 - \cos(k \, \Delta z' \cos \theta)] F(k).$$

$$(7.3.5)$$

The integral over θ yields:

$$d(\Delta z') = 8\pi^2 q_0^2 \int_0^\infty [1 - J_0(k \, \Delta z')]F(k)k \, dk, \quad (7.3.6)$$

which is identical to (6.3.10).

7.4. Internal-wave dominance for Φ and Λ

In this section we give some expressions for Φ, Λ, and related quantities from the internal-wave spectrum of sound-speed fluctuations, so that the reader may evaluate some specific cases, and gain some insight into the order of magnitude of various effects.

Local quantities

In the integral expressions for Λ (7.2.5) and Φ^2 (7.2.4); the local quantities $L_P(\theta, z)$ and L_V were introduced. Using the spectral function (3.6.5) for internal waves, we find (MZ76):

$$L_P(\theta, z) = \langle j^{-1} \rangle 4\pi^{-2}(n_0/\omega_i)B f_1(n\theta/\omega_i),$$

$$\langle j^{-1} \rangle = \sum_{j=1}^\infty j^{-1} H(j), \quad (7.4.1)$$

$$f_1(x) = (x^2 + 1)^{-1} + (x^2/2)$$
$$\times (x^2 + 1)^{-\frac{3}{2}} \ln \{[(x^2 + 1)^{\frac{1}{2}} + 1]/[(x^2 + 1)^{\frac{1}{2}} - 1]\},$$

$$L_V = (\pi j_* - 1)^{-1}(n_0/n)B, \quad (7.4.2)$$

$$L_H = \pi(n_0/\omega_i)B/\{8j_*[\ln(n/\omega_i) - 0.5]\}, \quad (7.4.3)$$

where $\langle j^{-1} \rangle = 0.439$ for $j_* = 3$ and $f_1(0) = 1$. It is of interest that $L_P(0, z)$ is *independent* of z and equal to 12.7 km. $L_P(\theta, z)$ diminishes rapidly with increasing θ (Fig. 7.5). $L_P(\theta, z)$ does not

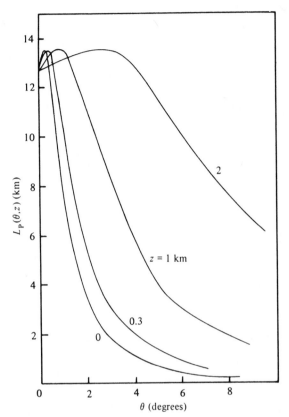

Fig. 7.5. The correlation length $L_P(\theta, z)$ as a function of orientation θ and depth z. Internal-wave dominance is assumed.

approach L_H at $\theta = 0$, nor L_V at large θ. This is because L_H and L_V are defined in terms of asymptotic limits at small displacements in the correlation function (6.1.11) while L_P is defined by an integral. Fig. 7.6 shows L_H and L_V as a function of depth.

Evaluation of Φ

We may evaluate Φ^2 from (7.2.4) given the above expression for $L_P(\theta, z)$, and the knowledge that for internal waves

$$\langle \mu^2(z) \rangle \approx \langle \mu_0^2 \rangle (n_0/n)(\partial_z C_P)^2/(\partial_z C_P)_0^2 = \langle \mu_0^2 \rangle (n/n_0)^3$$

$$(7.4.4)$$

where the first equality follows from (1.6.8) and the second follows

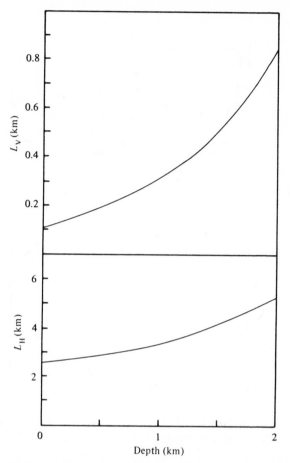

Fig. 7.6. The horizontal and vertical correlation lengths L_V and L_H as a function of depth, defined in terms of the asymptotic behavior of the sound-speed correlation function for small separations ((6.1.11) to (6.1.12)). Internal-wave dominance is assumed.

from the assumption of a constant T–S relation, as shown in (1.6.9). We present two examples here: more extensive treatment is given in Chapter 10.

The simplest case is a ray along the sound-channel axis. In that case $\theta = 0$, $z = z_1$, and,

$$\Phi^2 = q_0^2 R L_P(0) \langle \mu^2(z_1) \rangle \quad \text{(flat ray)}. \tag{7.4.5}$$

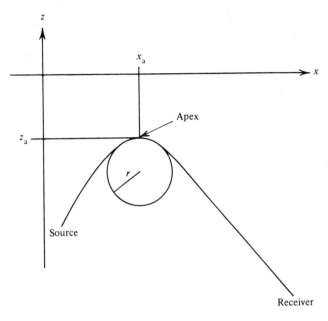

Fig. 7.7. A ray with a single upper turning point (apex).

A more complicated, but very useful case is a ray that makes a steep angle at the sound-channel axis. In that case, for long ranges, we may make an *apex approximation*. The majority of the contribution to Φ^2 comes from the region near the upper turning point of the ray, due to the fact that both $\langle \mu^2(z) \rangle$ and $L_P(\theta, z)$ have their maximum values at that point. Consider a ray (Fig. 7.7) with apex at position (x_a, z_a). Near the apex the equation of the ray is

$$z(x) = z_a - (1/2r)(x - x_a)^2,$$
$$\theta(x) = (x - x_a)/r, \tag{7.4.6}$$

where r is the radius of curvature of the ray.

The contribution of Φ^2 from an apex is then

$$\Phi_a^2 = q_0^2 \langle \mu^2(z_a) \rangle L_P(0)[\omega_i r/n(z_a)] \int_{-\infty}^{\infty} f_1(\alpha)\, d\alpha. \tag{7.4.7}$$

The integral can be evaluated (MZ76) as $\pi^2/2$. Let R_D be the range of a double loop for this steep ray. Then the number of upper turning points, each giving a contribution Φ_a^2, is R/R_D so that

$$\Phi^2 \approx q_0^2 R L_P(0) \langle \mu^2(z_a) \rangle \{(\pi^2/2)[\omega_i/n(z_a)](r/R_D)\} \quad \text{(steep ray)}.$$
$$(7.4.8)$$

Using (1.6.6) for our sound channel we may write (4.1.3)

$$r^{-1} \approx (2\varepsilon/B)\{\exp[2(z_a - z_1)/B] - 1\}, \qquad (7.4.9)$$
$$R_D \approx \pi B \varepsilon^{-\frac{1}{2}}, \qquad (7.4.10)$$

and also using (1.6.10) we may write both (7.4.5) and (7.4.8) in the form:

$$\Phi^2 = c(\theta_1)\langle \mu_0^2 \rangle L_P(0) q_0^2 R, \qquad (7.4.11)$$

where the coefficient c depends on the angle with which the ray crosses the sound-channel axis. Fig. 7.8 shows $c(\theta_1)$ as a function of angle, and illustrates that the value of Φ^2 is quite sensitive to the ray being considered, in addition to its more obvious dependences on range and acoustic wavelength. Of course, near $\theta_1 = 0$ (a flat ray) the contributions to Φ^2 come uniformly from each range increment, while for large θ_1 (steep rays with turning points near the ocean surface) the contributions come only from the regions close to the upper turning points, with one apex for each R_D increment in range. For convenience we have used (7.4.10) in evaluating R_D for both flat and steep rays, although we should point out that it is in error by about 35 per cent for steep rays.

Evaluation of Λ

The evaluation of Λ is similar to that of Φ except that two additional quantities must be considered; L_V and A^{-1}. The vertical correlation length, given by (7.4.2) is not a significant complication, but A^{-1} can be.

For flat rays (7.2.5) reduces to

$$\Lambda = \Phi^{-2} q_0^2 \langle \mu^2(z_1) \rangle L_P(0) L_V^{-2}(z_1) q_0^{-1} \int_0^R |A^{-1}| \, dx, \quad (7.4.12)$$

and A^{-1} is given by (7.1.12). For short ranges the result for a homogeneous, isotropic medium (6.2.3) is obtained. For *long range* the integral can be approximately evaluated as $4R|\pi^2 K \sin KR|^{-1}$ where $K = 2\pi/R_D$. Recalling the evaluation of Φ^2 for flat rays

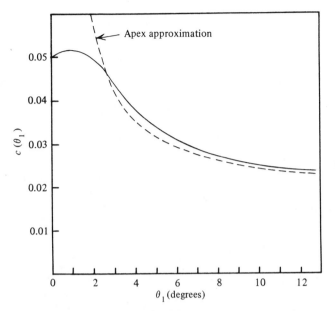

Fig. 7.8. The coefficient $c(\theta_1)$ necessary to evaluate the strength parameter Φ using (7.4.11) for a ray that crosses the sound-channel axis with angle θ_1. The solid curve has been obtained from a numerical integration of (7.2.4) using (7.4.1) and (7.4.4). The dashed curve has been obtained from the steep ray approximation (7.4.8).

(7.4.5) we arrive at:

$$\Lambda = 4 \left[\pi^2 q_0 L_V^2(z_1) K \sin KR\right]^{-1} \quad \text{(flat ray)}. \qquad (7.4.13)$$

Notice that Λ does not grow with range, but oscillates from infinite divergences when the receiver is placed at a caustic to a minimum value of

$$\Lambda_{\min} = (12/\pi^3) R_D / 6 q_0 L_V^2(z_1) \quad \text{(flat ray)}. \qquad (7.4.14)$$

For steep rays we again look to the *apex approximation*. This approximation will not be as good for Λ as in the case of Φ, because A^{-1} is zero at the apex (see (7.1.14)) so that we must retain higher terms in the expansion of A^{-1} around the apex. Nevertheless it still gives a reasonable picture of Λ. A rather simple, but tedious, calculation yields:

$$\Lambda = [R/6 q_0 L_V^2(z_a)](\overline{\theta^2}/\delta) \quad \text{(steep ray)} \qquad (7.4.15)$$

In the canonical sound channel (1.6.6) we find

$$\overline{\theta^2} = [32/(3\pi^3)]^{\frac{1}{2}} [\omega_i/n(z_a)](B/r)^{\frac{1}{2}} \qquad (7.4.16)$$

and the evaluation of $\overline{\theta^2}/\delta$ results in a reasonable approximation:

$$\Lambda \approx 0.023 \, R/6q_0 L_V^2(z_a) \quad \text{(steep ray).} \qquad (7.4.17)$$

It might appear that, for steep rays, the value of Λ in the ocean is more than an order of magnitude smaller than it would be in a homogeneous isotropic medium (6.2.3). It must be remembered however that L_V varies by an order of magnitude from the top to the bottom of a steep ray, and the small coefficient (0.023) is in fact representing a suitable average over the ray. For flat rays the value of Λ is on the average comparable to the result that would be obtained in a homogeneous isotropic ocean at a range of ~ 10 km. From the above examples the most important lesson to be learned is that the value of Λ is rather sensitive to small differences in the equilibrium ray under consideration.

Combinations of Λ and Φ

The boundaries in the Λ–Φ diagram involve combinations of Λ and Φ that will appear in other contexts, notably in coherence times, lengths and bandwidths (Chapter 8). It is thus useful to understand their meaning in terms of ocean fluctuation parameters. For a steep ray, using (7.4.11), we find

$$\Lambda\Phi^2 \approx (6 \times 10^{-4})\langle \mu_0^2 \rangle (R/L_V)^2 (L_P/\lambda), \qquad (7.4.18)$$

$$\Lambda\Phi \approx (6 \times 10^{-4})\langle \mu_0^2 \rangle^{\frac{1}{2}} (L_P/L_V)^{\frac{1}{2}} (R/L_V)^{\frac{3}{2}}, \qquad (7.4.19)$$

where L_V is evaluated at an apex, and L_P is evaluated at $\theta = 0$.

Thus $\Lambda\Phi^2$ is the product of three factors involving the variance of sound-speed fluctuations, the range in units of vertical correlation length, and the number of acoustic wavelengths in a horizontal correlation length. The three factors in $\Lambda\Phi$ are the rms sound-speed fluctuation, the ratio of horizontal to vertical correlation length (the anisotropy) and the range in units of vertical correlation length. $\Lambda\Phi$ is independent of acoustic frequency, while $\Lambda\Phi^2$ is proportional to acoustic frequency.

7.5. Evaluation of the phase-structure function

One may evaluate the phase-structure function in general using (7.3.2) and (7.3.3).

A complete expression for $d(\Delta z, \Delta y, \Delta t, \Delta\sigma)$ in terms of the internal-wave spectrum can be given in $\{\omega, j\}$ space:

$$d(\Delta z, \Delta y, \Delta t; z; \theta) = 4q_0^2 \sum_{j=1}^{\infty} \int_{\omega_L}^{n(\bar{z})} d\omega \, F(\omega, j, \bar{z}) k_y^{-1}$$

$$\cdot [1 - \cos(k_y \, \Delta y + k_V \, \Delta z) \cos \omega \, \Delta t] \qquad (7.5.1)$$

where

$$\omega_L^2 = \omega_i^2 + n^2 \tan^2 \theta, \qquad (7.5.2)$$

$$k_y = j\pi(\omega^2 - \omega_L^2)^{\frac{1}{2}}/n_0 B, \qquad (7.5.3)$$

$$k_V = j\pi n/n_0 B, \qquad (7.5.4)$$

and $F(\omega, j, \bar{z})$ is given by (3.6.5).

For the most part we are interested in the phase-structure function for small values of the separation variables. In this case a separation is possible. For example:

$$d(\Delta z, \Delta y, \Delta t; \theta; z) = d(\Delta z) + d(\Delta y) + d(\Delta t) \qquad (7.5.5)$$

where it is understood that all the functions on the right-hand side of (7.5.5) depend on the local depth z and unperturbed ray angle θ.

Vertical separations

When the power-law exponent of the sound-speed fluctuations is such that $p \approx 2$ the sum in (7.5.1) must be handled carefully. One cannot simply expand the cosine for small argument. Nevertheless a precise calculation yields a relatively simple answer:

$$d(\Delta z) = q_0^2 \langle \mu^2(z) \rangle L_P(\theta, z)(\Delta z/l_V)^2 [\ln(l_V/\Delta z) + \gamma_V] \qquad (7.5.6)$$

where the logarithm reflects the complications for a $p = 2$ spectrum.

The length l_V is given by

$$l_V = C_V L_V, \qquad (7.5.7)$$

$$C_V^2 = \sum_{j=1}^{\infty} (j^{-1})[j_*^2/(j^2 + j_*^2)] \approx \ln j_* \approx 1, \qquad (7.5.8)$$

and γ_V is given by

$$\gamma_V = \tfrac{3}{2} - \gamma + \ln(L_V/l_V), \qquad (7.5.9)$$

where γ is Euler's constant $(0.577\dots)$. Equation (7.5.6) is valid if $\Delta z \ll l_V$.

With $d(\Delta z)$ given above one càn integrate to obtain D:

$$D(\Delta z) = \int_0^R d(\Delta z')\,dx, \quad \Delta z' = \xi_1(x)\,\Delta z, \qquad (7.5.10)$$

where $\Delta z'$ is the ray separation at range x.

For short ranges we may set $\xi_1(x) = x/R$ and the various quantities in (7.5.6) can be considered as constants. Then

$$D_F(\Delta z) \approx \tfrac{1}{3}\,\Phi^2(\Delta z/l_V)^2[\ln(l_V/\Delta z) + \gamma_V + \tfrac{1}{3}] \qquad (7.5.11)$$

which may be compared with (6.3.16) in the homogeneous, isotropic case.

For a *single apex* the integral is dominated by the region around x_a:

$$D_A(\Delta z) \approx \Phi^2[\xi(x_a)\,\Delta z/l_V]^2\{\ln(l_V/\Delta z) + \gamma_V - \ln[\xi(x_a)]\}. \qquad (7.5.12)$$

For *long-range steep rays* we may combine (7.5.12) with the long-range approximation $\xi_1(x) = x\theta/R\theta(R)$, and we may approximate the sum over turning points as an integral, to obtain (roughly)

$$D(\Delta z) \approx \tfrac{1}{3}\Phi^2[\overline{\theta^2}/\theta^2(R)](\Delta z/l_V)^2[\ln(l_V/\Delta z) + \gamma_V], \qquad (7.5.13)$$

where l_V is to be evaluated at the turning-point depth, z_a, and $\overline{\theta^2}$ is defined in (7.4.16). A difficulty here is finding $\theta(R)$; a typical case would have $\theta(R) \approx \bar{\theta}$. One can see from all these evaluations that the result for a flat ray (7.5.11) gives a reasonable order of magnitude for D in all cases, and the result is not too different from a simple generalization of the homogeneous, isotropic case. However, quantitative estimates must take the particular ray into account, or errors of a factor of 5 to 10 could be incurred in extreme cases.

Horizontal separations

No simple analytic form for $d(\Delta y)$, analogous to (7.5.6), can be obtained, due to the complicated character of the integrals in (7.5.1). A numerical approximation, which is good to 30 per cent over most of the range of interest in the ocean (assuming internal-wave dominance) can be expressed as

$$d(\Delta y) = q_0^2 \langle \mu^2(z) \rangle L_P(\theta, z)(\Delta y/l_H)^2 \qquad (7.5.14)$$

where l_H is a fitted parameter. The value of l_H is approximately 3.7 km.

Time separations

For small time separations (and no spatial separations) the general formula (7.3.3) for d may be immediately approximated by expanding $\cos \omega \Delta t \approx 1 - \omega^2 \Delta t^2/2$, because the spectrum in ω is steep enough. The result of integrating over x is then:

$$D(\Delta t) = (\Delta t)^2 \dot{\Phi}^2. \tag{7.5.15}$$

In particular,

$$\dot{\Phi}^2/\Phi^2 = 2\omega_i^2 \ln (n_1/\omega_i) = 0.45 \text{ hour}^{-2} \quad \text{(flat ray),} \tag{7.5.16}$$

$$\dot{\Phi}^2/\Phi^2 = (4/\pi^2)\omega_i n_a (2\pi B/3r)^{\frac{1}{2}} \ln(n_a/\omega_i) \approx 3.3 \text{ hour}^{-2} \quad \text{(steep ray).} \tag{7.5.17}$$

CHAPTER 8

STATISTICS OF ACOUSTIC SIGNALS

The central question addressed in this book is how sound transmission is affected by propagation through a fluctuating ocean. We describe the ocean fluctuations statistically, hence their effects on sound signals must also be described statistically. Chapter 6 shows that in a homogeneous, isotropic medium the main regimes of signal-statistics behavior are regions in Λ–Φ space, where Λ is the diffraction parameter and Φ is the strength parameter. No less is true for the ocean medium. Within each regime the signal statistics are described in terms of ocean-fluctuation statistics (such as the phase-structure function). In this chapter we summarize results whose detailed justification is given in Part IV of this book.

A general description of acoustic signal statistics is given in §8.1. A physical picture of the different regimes in Λ–Φ space is given in §8.2 by means of a qualitative derivation. Ensemble averages at a point receiver (i.e. a single hydrophone) are treated in §8.3, which is followed by consideration of correlations between two or three different hydrophones separated either in time (§8.4), space (§8.5), or frequency (§8.6). Finally, pulse propagation to a single hydrophone is given separate treatment (§8.7), even though it is in principle contained in §8.6 as the Fourier transform of frequency separation.

8.1. Signal statistics and variables

For a medium without fluctuations or a sound channel the pressure field at a point in space–time is

$$p(\mathbf{x}, t, \sigma) = \exp\left[i(\mathbf{q} \cdot \mathbf{x} - \sigma t)\right]/(4\pi x) \qquad (8.1.1)$$

for a point source of unit strength at the origin with frequency σ and wavenumber q. Even without fluctuations, the sound channel modifies (8.1.1) drastically, to the point that we can only remove the time dependence explicitly:

$$p(\mathbf{x}, t, \sigma) = f_0(\mathbf{x}, \sigma) \exp(-\mathrm{i}\sigma t). \qquad (8.1.2)$$

When fluctuations are included,

$$p(\mathbf{x}, t, \sigma) = f(\mathbf{x}, t, \sigma) \exp(-\mathrm{i}\sigma t). \qquad (8.1.3)$$

We define the reduced wavefunction

$$\psi(\mathbf{x}, t, \sigma) = f(\mathbf{x}, t, \sigma)/f_0(\mathbf{x}, \sigma), \qquad (8.1.4)$$

which represents the effects of fluctuations in time and space, as a function of frequency, σ, the deterministic sound channel being contained to zeroth order in $f_0(\mathbf{x}, \sigma)$ and the time dependence being treated explicitly in (8.1.3). Hence in the absence of fluctuations $\psi(\mathbf{x}, t, \sigma) = 1$.

ψ is a complex random function whose statistical behavior we wish to discuss. Characterizing ψ is in general quite complicated, and for the most part we describe not ψ itself but rather some real function of ψ. For example:

$$\psi = \mathrm{Re}\,\psi + \mathrm{i}\mathrm{Im}\,\psi = A\,\mathrm{e}^{\mathrm{i}\phi} \qquad (8.1.5)$$

$$\ln \psi = \ln A + \mathrm{i}\phi = u + \mathrm{i}\phi = \tfrac{1}{2}\iota + \mathrm{i}\phi \qquad (8.1.6)$$

$$I = |\psi|^2 = |A|^2 = \mathrm{e}^{\iota} \qquad (8.1.7)$$

where

I = intensity,

ι = log-intensity,

A = amplitude,

u = log-amplitude,

ϕ = phase,

$\mathrm{Re}\,\psi$ = Cartesian real component,

$\mathrm{Im}\,\psi$ = Cartesian imaginary component.

To characterize ψ or any of the above functions we must give its probability density as a function of space, time, and frequency. The range coordinate (along the direction of propagation) is not an independent variable, so that we have, in general, a four-dimensional space. Every point in this four-dimensional space is *correlated* with every other point, and a complete description would require that all n-point correlation functions be given where $n = 1, 2, \ldots \infty$. In some cases (e.g. when the function is a Gaussian random function over the four-dimensional space) general formulas can be stated. However we are more commonly interested in a few salient numbers that roughly characterize signal statistics. The numbers we consider can be derived using only correlation functions for $n = 1, 2,$ and 3.

One-point functions

An example of a *one-point function* is the intensity, I, at a given point $(\mathbf{x}, t, \sigma)$. Under certain conditions, as will be shown later, the intensity has an exponential distribution (Rayleigh distribution in amplitude):

$$P(I) = a^{-1} \exp(-I/a). \tag{8.1.8}$$

From (8.1.8) one may find the expectation value of any function of I or of any moment of I; for example:

$$\langle I \rangle = a, \quad \langle I^n \rangle = n! a^n, \quad \langle \iota^2 \rangle - \langle \iota \rangle^2 = \pi^2/6, \tag{8.1.9}$$

all hold for an exponential distribution. The last result is a restatement of the well-known result that log-intensity fluctuations have an rms variation of 5.6 dB for a Rayleigh distribution (Dyer, 1970).

Another example of a one-point function is the phase ϕ. For a normal (Gaussian) distribution

$$P(\phi) = (2\pi b^2)^{-\frac{1}{2}} \exp[-\tfrac{1}{2}\phi^2/b^2], \tag{8.1.10}$$

in which case

$$\langle \phi \rangle = 0, \quad \langle \phi^2 \rangle = b^2. \tag{8.1.11}$$

Two-point functions

Two points in our (z, y, t, σ) space may be visualized as two hydrophones at different points in space, or as the same hydrophone recording at two different times, or at two different frequencies. The separation between the two points is characterized by *separation variables*:

$$\Delta z = z_2 - z_1 \quad \text{vertical separation,}$$

$$\Delta y = y_2 - y_1 \quad \text{horizontal separation (transverse to the direction of propagation),}$$

$$\Delta t = t_2 - t_1 \quad \text{time separation,}$$

$$\Delta \sigma = \sigma_2 - \sigma_1 \quad \text{frequency separation.} \tag{8.1.12}$$

It is always useful to perform a Fourier transform on ψ to obtain distributions in variables that are conjugate to the separation variables.

Separation variable	Conjugate variable	Physical meaning
Δz	$q_0 \theta_V$	θ_V = vertical arrival angle
Δy	$q_0 \theta_H$	θ_H = horizontal arrival angle
Δt	γ	Doppler shift
$\Delta \sigma$	τ	pulse time

We are interested in the correlations of ψ_2 and ψ_1 (or of any functions of ψ_2 and ψ_1) at all possible separations of the two points. Consider, as an example, the covariance of ψ. We know that in the limit of zero separation

$$\langle \psi_2^* \psi_1 \rangle \to \langle \psi_1^* \psi_1 \rangle \qquad (8.1.13)$$

and as separation goes to infinity

$$\langle \psi_2^* \psi_1 \rangle \to \langle \psi_2^* \rangle \langle \psi_1 \rangle. \qquad (8.1.14)$$

Fig. 8.1 illustrates the situation for a vertical separation, and also illustrates that a *characteristic scale* exists for each separation variable and its Fourier transform variable. In the following table we have divided the variables conjugate to spatial separations by q_0 to yield the more physically meaningful angles:

Separation variable	Characteristic scale	Conjugate variable	Characteristic scale
Δz	vertical coherence length	θ_V	rms vertical arrival angle
Δy	horizontal coherence length	θ_H	rms horizontal arrival angle
Δt	coherence time	γ	Doppler broadening
$\Delta \sigma$	coherent bandwidth	τ	pulse time extent

Three-point functions

To illustrate the importance of three-point functions, consider an example. A received signal, ψ, is recorded for 10^4 seconds, and a

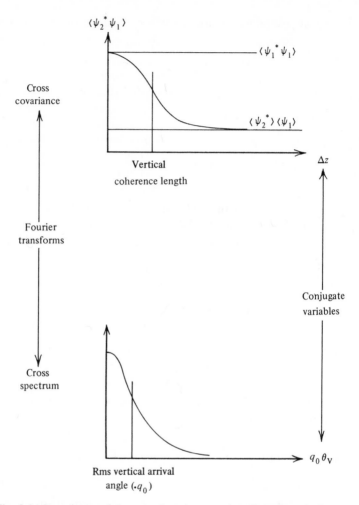

Fig. 8.1. Covariance of the wavefunction as a function of vertical separation, and its Fourier transform, the power spectrum. The characteristic scales in vertical separation and vertical arrival angle are indicated by the vertical lines.

Fourier transform of the signal is taken. The frequency resolution would be $\Delta \nu \approx 10^{-4}$ Hz. The shape of the power spectra might look like Fig. 8.2, with a *Doppler broadening* of 1 Hz. All of this information was obtained from two-point functions as described above. An important new question is one of stationarity: what is the *rate* at

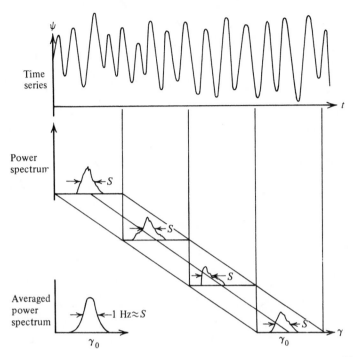

Fig. 8.2. Doppler broadening of a central carrier frequency γ_0. The broadening is dominated by *spread*, denoted by S.

which the Doppler broadening was achieved? That is, if the 10^4-second sample was split into one hundred 10^2-second samples (with an intrinsic resolution of 10^{-2} Hz), would each sample have a 1 Hz Doppler broadening, or would each sample have a much smaller broadening but centered at a different Doppler shift? In the first case we would refer to Doppler *spread* and in the second case to Doppler *wander* (Fig. 8.3). These terms *spread* and *wander* apply equally well to any of the conjugate variables, and the three-point functions are needed to distinguish where the data lie between these two extremes.

We will not develop extensive formalism for separating spread and wander until §8.7 where pulse time spread and pulse time wander are treated. The formalism described there can be applied directly to the other conjugate variables (arrival angle and Doppler shift).

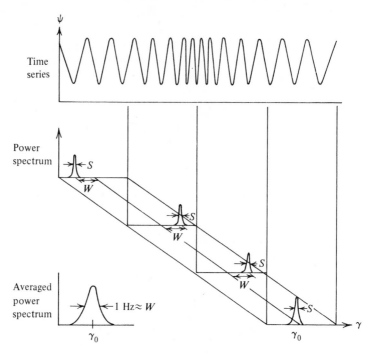

Fig. 8.3. Doppler broadening of a central carrier frequency γ_0. The broadening is dominated by *wander*, denoted by *W*.

8.2. Regimes in Λ–Φ space

The characteristic scales of *sound-speed* fluctuations are described by the strength parameter Φ and the diffraction parameter Λ (Chapter 7). We would like to characterize the behavior of *sound signals* in different regions of Λ–Φ space in a manner similar to that described in Chapter 6 for a homogeneous, isotropic medium.

All our treatments will start from the ray picture of wave propagation. In the unperturbed case there may be many rays connecting a source and receiver, but to a good approximation these separate deterministic rays are statistically independent, so we concentrate on the fluctuations in a single deterministic path.

We assume that we can calculate the behavior of the unperturbed ray; the problem in the presence of sound fluctuations is to *determine the new ray or rays.* By definition a ray is a curve with minimum

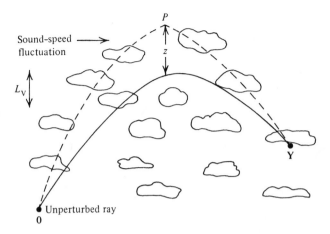

Fig. 8.4. A ray (dashed line) scattered at P, a distance z from the unperturbed ray.

(or maximum) acoustic path length from source to receiver. Therefore we must calculate the path length in the presence of sound fluctuations for all possible new curves and find the extrema. We will not try to accomplish that Herculean task here (it is attacked in Part IV), but rather we will pick only a small class of paths very close to the unperturbed ray to illustrate some general principles. Fig. 8.4 illustrates the geometry (cf. also Fig. 7.1), where the new path is separated from the unperturbed ray by a distance z at the midpoint, with the unperturbed ray followed from the origin ($\mathbf{0}$) to the point P, and from P to the receiver. The full path length from source to receiver can then be written

$$S(\mathbf{0}, \mathbf{Y}) = S_0(\mathbf{0},\mathbf{Y}) + \tfrac{1}{2}Az^2 - \int \mu \, ds \qquad (8.2.1)$$

where A is the phase curvature defined in Chapter 7 and the integral of μ is taken over the new path. The problem then is what value (or values) of z yield a minimum in S. At this point it is convenient to introduce L_V and to set $\xi = z/L_V$. We also define

$$u(\xi) = -\Phi^{-1}q_0 \int \mu \, ds \qquad (8.2.2)$$

so that $u(\xi)$ is a function whose magnitude is of order unity and which varies by order one when its argument ξ varies by order one.

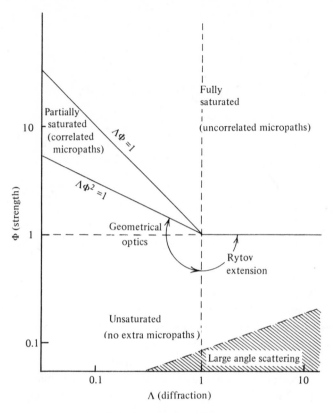

Fig. 8.5. Regions in Λ–Φ space.

We then find:

$$q_0 S(\mathbf{0}, \mathbf{Y}) \approx q_0 S_0(\mathbf{0}, \mathbf{Y}) + \tfrac{1}{2}\xi^2/\Lambda + \Phi u(\xi). \qquad (8.2.3)$$

Equation (8.2.3) has reduced the path length to a function of dimensionless variables ($u(\xi)$ and ξ) and the important parameters Φ and Λ. The regions of different sound fluctuation behavior can now be defined in terms of regions in Λ–Φ space (see Fig. 8.5).

The saturated regions

Suppose $\Phi\Lambda > 1$. The path length then has multiple minima because $u(\xi)$ is a random function whose derivatives are as likely to be negative as positive. It follows that for $\Phi\Lambda > 1$ there are several rays or micromultipaths connecting source to receiver. The question now arises as to whether or not these multiple rays are meaningful. In the wave theory two rays are distinct if the difference in path length satisfies $q_0|S_1 - S_2| \gtrsim 1$; i.e. if $S_1 - S_2$ is greater than about a quarter wavelength. For the micropaths $q_0|S_1 - S_2|$ is of order Φ and they are meaningful if $\Phi > 1$, but not if $\Phi < 1$.

Thus for propagation conditions such that $\Phi\Lambda > 1$ and $\Phi > 1$ each unperturbed ray will split up into a (fluctuating) number of micropaths. (On the average the number of micropaths is $\Lambda\Phi$.) The vertical spacing between these paths will typically be one unit in ξ or a distance L_V in z.

Actually the condition $\Phi\Lambda > 1$ is sufficient but not necessary for multipath. When the power spectrum of μ falls like an inverse power at high vertical wavenumbers, the smaller scales can produce micro-multipathing before the large scales become important. To see this, suppose that the power-law exponent for vertical wavenumber is $p = 2$. The length L_V is then to be interpreted as the inverse of a lower cutoff on the spectrum. Now delete all wavenumbers smaller than $1/(\lambda L_V)$ for some $0 < \lambda < 1$. The phase fluctuation Φ will become $\lambda\Phi$ and λL_V is the new basic scale. The quantity $\Phi\Lambda$ then becomes $\Phi\Lambda/\lambda$ which increases for small λ. If these micropaths are to be meaningful they must satisfy $\lambda\Phi > 1$ and the smallest meaningful value of λ is Φ^{-1}. Thus one sees that the small scales can produce micropaths if $\Phi^2\Lambda > 1$ (partially saturated). For a power law other than an inverse square the condition $\Phi^2\Lambda > 1$ is modified but the idea is the same.

Therefore, in the saturated regions the physics is being determined by the unperturbed ray breaking into many rays. In the fully saturated regime these new rays are spread vertically over a region larger than a correlation length L_V; in the partially saturated regime they are spread over a region smaller than L_V. The paths that are separated by less than L_V have some correlation with each other; the paths separated by more than a correlation length are statistically independent.

Quantitatively the above arguments may be summarized as follows, when $\Phi > 1$:

$\Phi\Lambda =$ *the total vertical range over which micropaths are spread in units of* L_V,

$\Phi^2\Lambda$ = *the rms phase difference between the extreme micropaths; also* = *total number of micropaths*,

Φ = *the number of micropaths within a vertical correlation length* L_V.

The unsaturated regions

In the *geometrical-optics* regime a single ray occurs at a very small value of ξ; i.e. the ray is displaced a small fraction of a vertical correlation length. In that case a perturbation expansion can be used in the variable ξ, resulting in a useful approximation for the sound-wave function, even if the phase fluctuation (Φ) is large. In the diffractive, unsaturated regime ($\Lambda > 1$, $\Phi < 1$) the best ray approximation is the unperturbed ray itself; the small scattered wave is not well described by a ray approximation, but because Φ is small a perturbation expansion can still be made, and the result turns out to be similar to the formulation in the geometrical-optics regime. This approximation is called the Rytov approximation, and yields

$$\psi = e^X \tag{8.2.4}$$

for the new wavefunction where X is an integral over the sound-speed fluctuations (see Chapter 10 for details).

Translation to range–frequency space

To determine in which region a particular sound-transmission experiment is operating, one must calculate Φ and Λ, which depend on the *sound frequency*, the *range* from source to receiver, and on *which unperturbed ray* is considered. Fig. 8.6 shows the regions translated into range–frequency space. This conversion has been made on the basis of (7.4.11) and (7.4.17) which are applicable to steep rays at long range, and which have been derived from the internal-wave model (Chapter 3); it should be used with considerable caution. For any given propagation geometry more reliable estimation of Φ and Λ can be obtained using the formulas of Chapter 7, generally requiring numerical integration.

8.3. One-point functions

The distribution of ψ itself at a given point constitutes the one-point problem. In this section we describe these statistics completely for

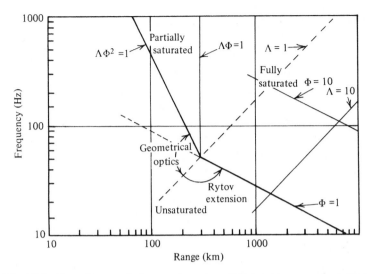

Fig. 8.6. Sound-transmission regions translated into range–frequency space. The translation from Fig. 8.5 has assumed a long-range, steep ray with scattering dominated by internal waves according to the known spectrum (Chapter 3).

points in Λ–Φ space far from any boundary. One boundary ($\Lambda = 1$ within the unsaturated region) can be treated also.

The unsaturated region

In the presence of ocean fluctuations only one perturbed ray exists in the unsaturated region. One then expects that phase and amplitude are simple quantities, and this is borne out by the fact that (8.2.4) is valid, where

$$u = \mathrm{Re}\,X, \quad \phi = \mathrm{Im}\,X, \tag{8.3.1}$$

are linear functions of μ and hence have Gaussian probability distributions. In that case A has a log-normal distribution. The ϕ and u statistics are described completely by giving their variances. First we give the rather simple results far from all boundaries:

$$\langle \phi^2 \rangle = \Phi^2, \quad \langle u^2 \rangle \approx \tfrac{1}{4}\Lambda\Phi^2, \quad \text{for } \Lambda \ll 1 \tag{8.3.2}$$

$$\langle \phi^2 \rangle = \langle u^2 \rangle = \tfrac{1}{2}\Phi^2, \quad \text{for } \Lambda \gg 1 \tag{8.3.3}$$

for small and large diffraction respectively. In the former case the phase is more strongly perturbed than the log-amplitude, in the latter case the phase and log-amplitude have equal variances.

The general result, valid even on the boundary $\Lambda = 1$, requires an ocean-fluctuation function (Chapter 10) which we can schematically indicate as

$$\langle X^2 \rangle = -\Phi^2 \langle \exp{(ik_z^2 R_F^2/2\pi)} \rangle \tag{8.3.4}$$

where the averaging is over the spectrum of vertical wavenumber and Fresnel-zone radius along the unperturbed ray. The general result for the variances is

$$\langle \phi^2 \rangle = \tfrac{1}{2}(\Phi^2 - \text{Re}\,\langle X^2 \rangle), \quad \langle u^2 \rangle = \tfrac{1}{2}(\Phi^2 + \text{Re}\,\langle X^2 \rangle). \tag{8.3.5}$$

We can see that for large Λ, $\langle X^2 \rangle \approx 0$, and hence (8.3.3) follows. For small Λ, $\langle X^2 \rangle \approx -\Phi^2$ and hence $\langle \phi^2 \rangle \approx \Phi^2$, and $\langle u^2 \rangle \approx 0$. The log-amplitude fluctuations in the geometrical-optics region are thus of second order, and a careful treatment yields (8.3.2), where the exact coefficient of $\Lambda\Phi^2$ depends on details of the ocean-fluctuation spectrum (Chapter 10) but is of order unity.

The saturated regions

Ocean fluctuations in the saturated regions cause many perturbed rays to appear in the place of one unperturbed ray. As a result the overall ψ is the sum of many random contributions. In this case the central-limit theorem implies that Re ψ and Im ψ are both Gaussian random variables with zero mean and variance

$$\langle (\text{Re}\,\psi)^2 \rangle = \langle (\text{Im}\,\psi)^2 \rangle = \tfrac{1}{2}\langle |\psi|^2 \rangle = \tfrac{1}{2}, \tag{8.3.6}$$

where the last equality follows because $\langle |\psi|^2 \rangle = 1$ by conservation of energy. (In a region of rapidly-varying *unperturbed* energy, our equations break down.)

Phase in this region is very difficult to analyze, because it is a result of an addition of many paths. As a result the phase cannot be defined using one-point functions. A measurement of phase for one member of an ensemble cannot be related to a measurement of another member. (The phase will become useful when we consider two-point functions.)

The probability distribution of intensity (or amplitude) is determined from (8.3.6) and the knowledge that Re ψ and Im ψ are

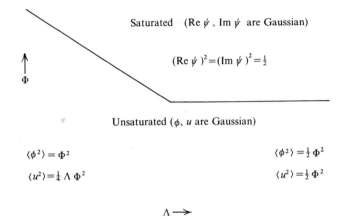

Fig. 8.7. One-point-function behavior summarized in Λ–Φ space.

Gaussian with zero mean. The result is well known:

$$P(I) = \exp(-I), \quad P(A) = 2A \exp(-A^2). \qquad (8.3.7)$$

The amplitude has a Rayleigh distribution.

Λ–Φ space

A summary of the above results far from boundaries is given in Fig. 8.7. If we draw the unperturbed phasor (complex amplitude) in each regime as an arrow of length unity and zero phase, then the ocean fluctuations result in a probability distribution for the perturbed ψ to lie at some point in the complex plane. A contour of equal probability is drawn in Fig. 8.8 for each regime to illustrate the shape of the probability distribution.

Boundaries

At the boundary $\Lambda = 1$ in the unsaturated region ϕ and u are still Gaussian with variances given by (8.3.5). (For further details see Chapter 10.) However the statistics at the boundary between the unsaturated and saturated region have not yet been understood. To illustrate the problem consider the boundary $\Phi = 1$ when $\Lambda \gg 1$. From above, Re ψ and Im ψ are Gaussian, which would imply that the amplitude statistics are Rice–Nakagami at the boundary. Let Re ψ be Gaussian with a displaced mean m, and both Re ψ and Im ψ

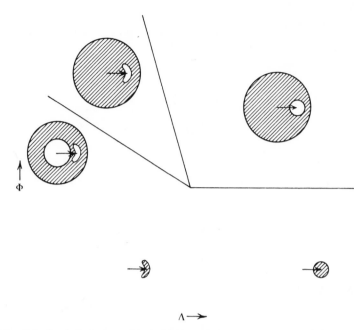

Fig. 8.8. Statistical-ensemble behavior of the wavefunction phasor. The arrow indicates the wavefunction phasor unperturbed by any sound-speed fluctuations. The outside boundary of the shaded region indicates a line of constant probability of finding the perturbed phasor. There is a small probability of finding the phasor outside the outer boundary of the shaded region. The inner unshaded region indicates the most likely place to find the phasor a short period from the time when the phasor happens to have its unperturbed value. In this figure, as in Figs. 8.9 to 8.15, the boundaries between regions are indicated schematically; the slopes of boundary lines are not to scale.

have variances of $\frac{1}{2}\sigma^2$;

$$P(A) = (2A/\sigma^2) \exp\left[-(A^2 + m^2)/\sigma^2\right] I_0(2Am/\sigma^2) \quad (8.3.8)$$

where I_0 is a modified Bessel function (Beckmann, 1967; Norton *et al.* 1955). On the other hand, from below one might expect that u and ϕ remain Gaussian which would imply that A is log-normal. It is interesting to note however that (8.3.8) has the correct behavior at both large Φ and small Φ; that is, the assumption that Re ψ and Im ψ are Gaussian implies the log-normal and Rayleigh distribution as limits for small and large fluctuations, respectively. The mean m

can be derived as (see Chapter 11)

$$m \equiv \langle \operatorname{Re} \psi \rangle = \exp\left(-\tfrac{1}{2}\Phi^2\right), \qquad (8.3.9)$$

but only the limits of the variance have been calculated. The other saturated–unsaturated boundary ($\Lambda\Phi^2 = 1$, $\Lambda < 1$) is even less understood.

8.4. Time separations

Suppose that a source emits sound of frequency σ and a receiver at **x** records the complex pressure at two different times separated by Δt. Various combinations of these two ψs are *two-point functions*. We will describe their probability distributions.

The unsaturated regions

Phase and log-amplitude are the useful quantities in the unsaturated regions. It can be shown that *phase difference* and *log-amplitude difference* are Gaussian random variables with zero mean. The variances in the geometrical-optics region ($\Lambda \ll 1$) are

$$\langle [\phi(\Delta t) - \phi(0)]^2 \rangle = D(\Delta t), \qquad (8.4.1)$$

$$\langle [u(\Delta t) - u(0)]^2 \rangle \approx 0, \qquad (8.4.2)$$

and in the diffractive region ($\Lambda \gg 1$)

$$\langle [\phi(\Delta t) - \phi(0)]^2 \rangle = \tfrac{1}{2}D(\Delta t), \qquad (8.4.3)$$

$$\langle [u(\Delta t) - u(0)]^2 \rangle = \tfrac{1}{2}D(\Delta t), \qquad (8.4.4)$$

where $D(\Delta t)$ is the phase-structure function defined in Chapter 7. From the above results nearly everything of interest can be derived (Chapter 10). For example the spectrum of phase difference, corresponding to the Doppler-shift spectrum, would be related to the Fourier transform of $D(\Delta t)$.

The cross-correlation of the wavefunction at different times can be expressed as (see Chapter 11)

$$\langle \psi^*(\Delta t)\psi(0) \rangle = \exp\left[-\tfrac{1}{2}D(\Delta t)\right]. \qquad (8.4.5)$$

The saturated regions

In the saturated regions the total signal is made up of a sum of micromultipaths. In that case it can be shown that the cross-correlation of the wavefunction is again given by (8.4.5), which is therefore *a universal result valid in all regimes.*

A separation between the partially and fully saturated regimes occurs when intensity correlations are considered. In the fully saturated region the interfering multipaths are uncorrelated and $\psi(t)$ and $\psi(0)$ constitute two functions obeying two-point Gaussian statistics; then the intensity correlations are (see Chapter 9):

$$\langle I(\Delta t)I(0)\rangle = 1 + \exp\left[-D(\Delta t)\right]. \tag{8.4.6}$$

On the other hand, in the partially saturated region the multipaths are correlated, so that one may think of the full wavefunction as being formed from a product of two functions:

$$\psi(\Delta t) = \exp\left[i\phi_0(\Delta t)\right]\chi(\Delta t), \tag{8.4.7}$$

where $\chi(\Delta t)$ represents the effect of interfering multipaths, and $\phi_0(\Delta t)$ represents an overall phase that is common to all the paths because they are correlated. Under these conditions the intensity correlation falls much more slowly because $\phi_0(\Delta t)$ cancels in the intensity. We may write:

$$\langle I(\Delta t)I(0)\rangle = 1 + K(\Delta t/t_0); \quad t_0 = \pi^2(6\Lambda\Phi^2 \ln \Phi)^{-1}(\Phi/\dot{\Phi}). \tag{8.4.8}$$

The method of calculating $K(\Delta t/t_0)$ is given in Appendix A.

Λ–Φ *space*

The above results are summarized in Fig. 8.9. It is useful to visualize how the wavefunction ψ wanders in time to eventually fill up the contours of probability drawn in Fig. 8.8. In the fully saturated region and in the diffractive unsaturated region Re ψ and Im ψ are uncorrelated and execute a random walk to fill up their ultimate circle (Fig. 8.10). In the geometrical-optics region the phase wanders according to two-point Gaussian statistics, while the log-amplitude has small variations. If $\Phi > 1$ an annulus is gradually filled. In the partially saturated region the full circle is eventually filled, but the motion in phase (8.4.5) is more rapid than the motion in magnitude (8.4.8) so that a spiral form is taken. The qualitative connection of these various forms across the boundaries is evident from Fig. 8.10.

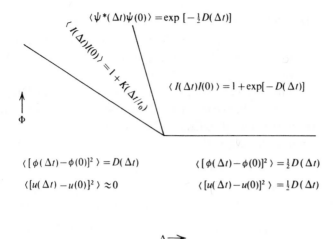

$$\langle \psi^*(\Delta t)\psi(0)\rangle = \exp\left[-\tfrac{1}{2}D(\Delta t)\right]$$

$$\langle I(\Delta t)I(0)\rangle = 1 + Kf(\Delta t/t_o)$$

$$\langle I(\Delta t)I(0)\rangle = 1 + \exp[-D(\Delta t)]$$

$$\langle[\phi(\Delta t)-\phi(0)]^2\rangle = D(\Delta t)$$

$$\langle[u(\Delta t)-u(0)]^2\rangle \approx 0$$

$$\langle[\phi(\Delta t)-\phi(0)]^2\rangle = \tfrac{1}{2}D(\Delta t)$$

$$\langle[u(\Delta t)-u(0)]^2\rangle = \tfrac{1}{2}D(\Delta t)$$

$\Lambda \rightarrow$

Fig. 8.9. Time-separation behavior summarized in Λ–Φ space.

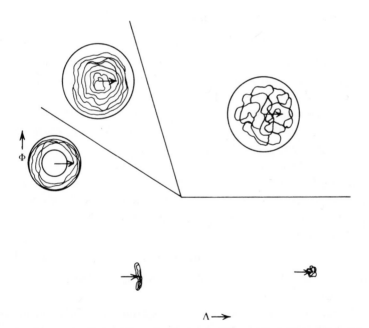

$\Lambda \rightarrow$

Fig. 8.10. Time-separation behavior of the wavefunction phasor. The horizontal arrows represent the unperturbed phasor. The curves indicate a typical time history of a phasor in the different regions of Λ–Φ space.

Coherence times

As discussed in § 8.1, the main feature to be determined from a two-point function is a characteristic scale, in this case coherence time. Because (8.4.5) is valid everywhere, it is clear that the coherence time of greatest importance is the time scale for $D(\Delta t)$ to make a significant change in $\exp\left[-\frac{1}{2}D(\Delta t)\right]$. However we know that the maximum value of $D(\Delta t)$ is $2\Phi^2$. For small Φ^2 the natural definition of a characteristic scale is

$$D(T_c) = \Phi^2, \quad \Phi < 1. \tag{8.4.9}$$

Using (7.5.15)

$$T_c = \Phi/\dot{\Phi} \approx 1 \text{ hour}, \quad \Phi < 1 \tag{8.4.10}$$

where the numerical evaluation has used the internal-wave result, (7.5.16) or (7.5.17).

For large Φ the exponential is significantly affected when

$$D(T_c) = 1, \Phi > 1, \tag{8.4.11}$$

and again using (7.5.15) we have

$$T_c = \Phi^{-1}(\Phi/\dot{\Phi}) \approx (1 \text{ hour})/\Phi, \quad \Phi > 1. \tag{8.4.12}$$

The *intensity correlation* gives a similar result for coherence times except in the *partially saturated region*. In that case the characteristic time t_0 of (8.4.8) is involved.

An important caveat needs to be added here. The intensity coherence time (8.4.8) is calculated from the intrinsic internal-wave time dependence. Another time dependence has been neglected; that caused by a convecting current in the area between source and receiver. One may estimate the coherence time from a horizontal current of speed v_c as $T_H = d_H/\sqrt{3}v_c$ where d_H is the horizontal coherence length for a point source (see § 8.5). Thus for a current with speed such that $T_H < t_0$,

$$v_c > d_H/\sqrt{3}T_H, \tag{8.4.13}$$

the effect of the current will dominate and $K(\Delta t/t_0)$ becomes irrelevant.

If there were no internal-wave intrinsic time dependence at all, so that both phase and amplitude of ψ varied only because of a convecting current, the behavior in time would be identical to the behavior in space (§ 8.5) scaled such that $\Delta y = v_c \Delta t$.

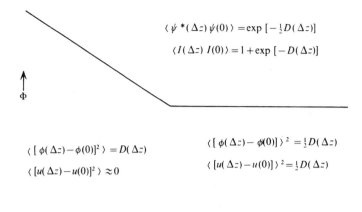

$$\langle \psi^*(\Delta z)\,\psi(0) \rangle = \exp\left[-\tfrac{1}{2}D(\Delta z)\right]$$

$$\langle I(\Delta z)\,I(0) \rangle = 1 + \exp\left[-D(\Delta z)\right]$$

$$\langle [\phi(\Delta z) - \phi(0)]^2 \rangle = D(\Delta z)$$

$$\langle [u(\Delta z) - u(0)]^2 \rangle \approx 0$$

$$\langle [\phi(\Delta z) - \phi(0)] \rangle^2 = \tfrac{1}{2}D(\Delta z)$$

$$\langle [u(\Delta z) - u(0)] \rangle^2 = \tfrac{1}{2}D(\Delta z)$$

$$\Lambda \longrightarrow$$

Fig. 8.11. Space-separation behavior summarized in Λ–Φ space.

8.5. Spatial separations

Suppose that a source emits sound of frequency σ, and two receivers, one at **x** and the other separated from **x** (by a vertical distance Δz or a horizontal distance Δy), record their complex pressures at the same time. Various combinations of these two ψs are two-point functions. We describe their probability distributions below.

This problem is simpler than that of time separation. The results are identical to the time separation case except in the partially saturated region, where the results are, in the spatial case, equivalent to the fully saturated results.

Λ–Φ space

Figure 8.11 lists the important formulas for the spatial separation case where the separation is vertical (replace Δz by Δy for the horizontal case). Fig. 8.12 shows the way the wavefunction ψ develops as one moves vertically (or horizontally) away from an arbitrary reception point.

Coherence lengths

Again the most basic equation for coherence is, by analogy with (8.4.5):

$$\langle \psi^*(\Delta y)\psi(0) \rangle = \exp\left[-\tfrac{1}{2}D(\Delta y)\right]. \tag{8.5.1}$$

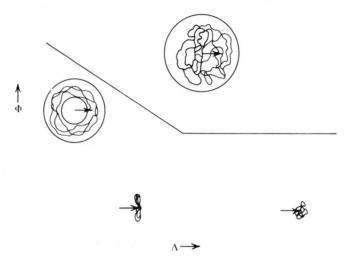

Fig. 8.12. Space-separation behavior of the wavefunction phasor. The curves indicate a typical behavior of the phasor, where the parameter along the curve is spatial separation.

We may define a horizontal coherence length, d_H, such that the argument of the exponential equals unity, applicable only to the case of large Φ. Using (7.5.14) we can approximately evaluate d_H as

$$d_H \approx \sqrt{6} l_H/\Phi, \quad \Phi > 1. \tag{8.5.2}$$

The analysis in Δz must begin from a formula for $D(\Delta z)$ appropriate to the particular ray. For example (7.5.11) would lead to a vertical correlation length

$$d_V \approx \sqrt{6} l_V/\Phi \ln \Phi, \quad \Phi > 1. \tag{8.5.3}$$

8.6. Frequency separations

Suppose that a source emits sound simultaneously at two different frequencies σ_1 and σ_2; let $\Delta\sigma = \sigma_2 - \sigma_1$ and $\sigma = \frac{1}{2}(\sigma_1 + \sigma_2)$. A receiver at **x** records the complex pressure at both frequencies, yielding two wavefunctions $\psi(\sigma_1)$ and $\psi(\sigma_2)$.

The unsaturated regions

Where u and ϕ are Gaussian random variables (the unsaturated regions) the frequency dependence is simple. We note that both u

and ϕ are linear functions of σ. Therefore we can write

$$\phi(\sigma_2) - \phi(\sigma_1) = (\Delta\sigma/\sigma)\phi(\sigma) \qquad (8.6.1)$$

$$u(\sigma_2) - u(\sigma_1) = (\Delta\sigma/\sigma)u(\sigma) \qquad (8.6.2)$$

so that phase difference and log-amplitude difference are also Gaussian random variables, with variance

$$\langle[\phi(\sigma_2) - \phi(\sigma_1)]^2\rangle = (\Delta\sigma/\sigma)^2\langle\phi^2(\sigma)\rangle, \qquad (8.6.3)$$

$$\langle[u(\sigma_2) - u(\sigma_1)]^2\rangle = (\Delta\sigma/\sigma)^2\langle u^2(\sigma)\rangle. \qquad (8.6.4)$$

We know $\langle\phi^2\rangle$ and $\langle u^2\rangle$ from § 8.3, so all information is given.

The distribution of the complex coherence is more complicated. We may write from (8.2.4) and the knowledge that X is linear in σ:

$$\psi^*(\sigma_2)\psi(\sigma_1) = \exp[2\,\mathrm{Re}\,X + i(\Delta\sigma/\sigma)\,\mathrm{Im}\,X]. \qquad (8.6.5)$$

We know that $\mathrm{Re}\,X$ and $\mathrm{Im}\,X$ are Gaussian random variables with zero mean:

$$\langle\psi^*(\sigma_2)\psi(\sigma_1)\rangle = \exp[2\langle(\mathrm{Re}\,X)^2\rangle - \tfrac{1}{2}(\Delta\sigma/\sigma)^2\langle(\mathrm{Im}\,X)^2\rangle$$

$$+ 2i(\Delta\sigma/\sigma)\langle\mathrm{Re}\,X\,\mathrm{Im}\,X\rangle]. \qquad (8.6.6)$$

In general the argument of the exponential is difficult to evaluate (Chapter 10). However, far from the $\Lambda = 1$ boundary we may arrive at some useful results. For $\Lambda \gg 1$ (diffraction) we know that $\mathrm{Re}\,X$ and $\mathrm{Im}\,X$ are uncorrelated with equal variance. Normalizing to the $\Delta\sigma = 0$ point and making use of (8.3.3) we find

$$\langle\psi^*(\sigma_2)\psi(\sigma_1)\rangle = \exp[-\tfrac{1}{2}(\Delta\sigma/\sigma)^2\langle\phi^2\rangle]. \qquad (8.6.7)$$

In the geometrical-optics region ($\Lambda \ll 1$), $\mathrm{Re}\,X$ and $\mathrm{Im}\,X$ are correlated, but terms involving the expectation of $\mathrm{Re}\,X$ are proportional to $\Lambda\Phi^2$, so that (8.6.7) still holds if $\Lambda \ll \Delta\sigma/\sigma$. If $\Lambda \gtrsim \Delta\sigma/\sigma$ an imaginary term appears in the argument of the exponential of the form $(\Delta\sigma/\sigma)\Lambda\Phi^2$.

Taking (8.6.7) as approximately valid throughout the unsaturated region, we have obtained the answer in useful form, since $\langle\phi^2\rangle$ is known from § 8.3.

The saturated regions

In the saturated region the signal is the sum of many micropaths, each of which satisfies geometrical optics in the sense that the stationary-phase approximation is good. Therefore for each micropath

$$\langle \psi_i^* (\sigma_2)\psi_i(\sigma_1)\rangle = \exp\left[-\tfrac{1}{2}(\Delta\sigma/\sigma)^2\Phi^2\right]. \qquad (8.6.8)$$

However the full signal is the sum of many micropaths. With a variable frequency one can form pulses and the micropaths can be separated by time-of-arrival. A frequency correlation must therefore explicitly contain information about the micropaths. There are no micropaths in first-order geometrical optics, so that (8.6.8) cannot be valid. We can write the correct formula in the form

$$\langle \psi^*(\sigma_2)\psi(\sigma_1)\rangle = \exp\left[-\tfrac{1}{2}(\Delta\sigma/\sigma)^2\Phi^2\right]Q(\Delta\sigma/\eta_0) \qquad (8.6.9)$$

where the Q factor is a result of micropath interference, and decreases significantly when $\Delta\sigma$ reaches a value comparable to the characteristic scale η_0. The value of η_0 is (see Chapter 12),

$$\eta_0 = \sigma(\pi^2/6)(l_V/L_V)^2(\Lambda\Phi^2\ln\Phi)^{-1}, \qquad (8.6.10)$$

where l_V (the scale in the phase-structure function) and L_V (the scale in the sound-speed correlation function) are evaluated in Chapter 7 for internal waves.

The quantity η_0 is roughly independent of acoustic frequency (since $\Lambda\Phi^2\approx\sigma$) and may be regarded as the bandwidth of the micropath effects. The function Q depends on the deterministic sound channel and the source–receiver geometry, and it is a complex function; its calculation is described in Appendix B.

Comparing the two factors in (8.6.9) we see that the Q factor dominates in the fully saturated region, whereas the exponential factor dominates in the partially saturated region.

Fig. 8.13 summarizes the formulas obtained for frequency correlation.

Coherent bandwidth, η_W

Throughout the Λ–Φ diagram equation (8.6.8) is approximately valid except in the fully saturated region. Wherever $\Phi < 1$ there is no limitation to bandwidth from (8.6.8) since the maximum bandwidth that can theoretically be obtained with a perfect transmission channel is $(\Delta\sigma/\sigma)\sim 1$. When $\Phi > 1$ we have a bandwidth limitation of

$$\eta_W/\sigma \approx 1/\Phi, \qquad (8.6.11)$$

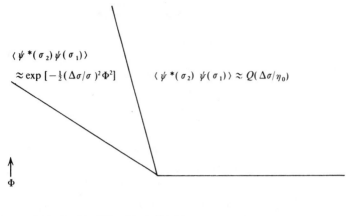

$$\langle \psi^*(\sigma_2)\psi(\sigma_1)\rangle$$
$$\approx \exp\left[-\tfrac{1}{2}(\Delta\sigma/\sigma)^2\Phi^2\right] \qquad \langle \psi^*(\sigma_2)\,\psi(\sigma_1)\rangle \approx Q(\Delta\sigma/\eta_0)$$

Φ

$$\langle[\phi(\sigma_2)-\phi(\sigma_1)]^2\rangle = (\Delta\sigma/\sigma)^2\langle\phi^2\rangle$$

$$\langle[u(\sigma_2)-u(\sigma_1)]^2\rangle = (\Delta\sigma/\sigma)^2\langle u^2\rangle$$

$\Lambda\longrightarrow$

Fig. 8.13. Frequency-separation behavior of the wavefunction phasor summarized in Λ–Φ space.

in the unsaturated or partially saturated region. In the fully saturated region the bandwidth available is limited by Q so that $\eta_{\mathrm{w}} \approx \eta_0$;

$$\eta_{\mathrm{w}} \approx \sigma(\pi^2/6)(l_{\mathrm{V}}/L_{\mathrm{V}})^2(\Lambda\Phi^2\ln\Phi)^{-1}. \qquad (8.6.12)$$

As an example we may evaluate (8.6.12) for steep rays at long range using the results of Chapter 7. The result is

$$\eta_{\mathrm{w}} \approx 8\ \mathrm{Hz}\ (1000\ \mathrm{km}/R)^2. \qquad (8.6.13)$$

This bandwidth assumes the fluctuation in a single deterministic ray is recorded. A further limitation would be introduced when many deterministic rays exist from source to receiver (§ 8.7), and they are not separated in some manner.

Time–bandwidth product

In the saturated region it is of interest to calculate the time–bandwidth product, viewing the ocean as a coherent communications channel. The result of combining the coherence time

(8.4.12) with the coherent bandwidth (8.6.12) yields

$$T_c \eta_W \approx \sigma \Phi^{-1} (\Phi/\dot{\Phi})(\Lambda \Phi^2 \ln \Phi). \qquad (8.6.14)$$

Expressing these quantities in terms of a spectrum of internal waves by using the results of Chapter 7, we may write:

$$T_c \eta_W \approx 10^4 \langle \mu_0^2 \rangle^{-\frac{3}{2}} (\lambda/R)(C/v_I)(L_V/L_P)^2 (L_P/R)^{\frac{3}{2}}, \qquad (8.6.15)$$

where $v_I = L_P \dot{\Phi}/\Phi$ is related to the speed of internal waves and we have evaluated the constant for a steep ray. Thus the large ratio of sound speed to internal-wave speed, and the long horizontal correlation length, favor a large time–bandwidth product, while the anisotropy (L_V/L_P) lowers it. A typical value, for $R = 300$ km and $\sigma = 1000$ Hz is

$$T_c \eta_W \approx 3000 \text{ cycles.} \qquad (8.6.16)$$

8.7. Pulse propagation

Suppose a narrow pulse is transmitted from source to receiver. Since the speed of sound (1500 m s^{-1}) is considerably greater than the speed of ocean fluctuations $(v \sim 2 \text{ m s}^{-1}$ for internal waves) we can assume that the ocean is frozen during the pulse transmission. Since any pulse can be made of a superposition of many frequencies, everything in this section could be derived from the results of § 8.6. However, experiments of this type are so common that explicit discussion is warranted (Ellinthorpe and Nuttall, 1965).

Let τ measure the time during pulse transmission with $\tau = 0$ at the instant of emission for an extremely sharp pulse. Then the received wave function as a function of τ is $h(\tau)$, where $h(\tau)$ is sometimes called the *impulse response*. It is useful to define the Fourier transform of $h(\tau)$

$$H(\sigma) = \int_{-\infty}^{\infty} h(\tau) \exp(-i\sigma\tau) \, d\tau, \qquad (8.7.1)$$

where $H(\sigma)$ is sometimes called the *instantaneous transfer function*. $H(\sigma)$ is in fact the received pressure for a signal transmitted at constant frequency σ. It is, therefore, identical to $\psi(\sigma)$ used in § 8.6.

Consider the intensity received at time τ in this pulsed experiment. We may write

$$I(\tau) = h^*(\tau)h(\tau)$$
$$= (4\pi S)^{-1} \int_{-S}^{S} d\sigma \int_{-\infty}^{\infty} d\Delta\sigma H^*(\sigma_2)H(\sigma_1) \exp{(i\Delta\sigma\tau)}. \qquad (8.7.2)$$

The limit $S \to \infty$ is taken to represent a δ-function input. The average of $I(\tau)$ over an ensemble is thus expressed directly in terms of the average over the two-point function $\langle \psi^*(\sigma_2)\psi(\sigma_1) \rangle$ discussed in §8.6. For example, in the saturated regions

$$\langle I(\tau) \rangle = (4\pi S)^{-1} \int_{-S}^{S} d\sigma \int d\Delta\sigma$$
$$\times \exp{[-\tfrac{1}{2}(\Delta\sigma/\sigma)^2\Phi^2]} Q(\Delta\sigma/\eta_0) \exp{(i\Delta\sigma\tau)}. \qquad (8.7.3)$$

The important *characteristic scale* of pulse time τ, called the *pulse time extent*, will be given by the width of $\langle I(\tau) \rangle$. Note that Φ/σ is independent of σ, and can be called the rms variation in acoustic travel time, T_{rms}. Thus from (8.7.3) the pulse time extent is T_{rms} or $(\pi\eta_0)^{-1}$, whichever is larger. In other words the term in the integrand with the smallest bandwidth dominates the pulse extension.

Spread and wander

The pulse time extent may be a consequence of the fact that $h(\tau)$ is *spread* in τ for each realization of the ensemble, or it may be that $h(\tau)$ is very narrow, but *wanders* to different τ from one member of the ensemble to another. To distinguish these cases a convenient function is the lagged intensity covariance

$$F(\tau') = \int_{-\infty}^{\infty} I(\tau + \tau')I(\tau)\,d\tau, \qquad (8.7.4)$$

which is in fact a three-point function. (One realizes that $I(\tau)$ is a two-point function since τ is a separation variable (§8.1), and $I(\tau + \tau')$ brings a third point at $\tau + \tau'$ into the expression.) Expressing (8.7.4) in Fourier transforms, we have

$$F(\tau') = (8\pi S^2)^{-1} \int_{-S}^{S} d\sigma \int_{-S}^{S} d\sigma' \int d\Delta\sigma\, H(\sigma + \Delta\sigma/2)H^*(\sigma - \Delta\sigma/2)$$
$$\times H(\sigma' - \Delta\sigma/2)H^*(\sigma' + \Delta\sigma/2) \exp{(i\Delta\sigma\tau')}.$$
$$\qquad (8.7.5)$$

The calculation of this integral looks forbidding, but in fact it can be done in several important cases. Suppose that we are in the geometrical-optics region, so that only one ray exists from source to receiver (that is, no micropaths and no deterministic multipath). If that ray has acoustic travel time T, then

$$H(\sigma) = \exp(-i\sigma T). \qquad (8.7.6)$$

In that case the integral reduces immediately to

$$F(\tau') = \delta(\tau') \qquad (8.7.7)$$

so that *no spread* in τ is observed.

On the other hand, one can show that

$$\langle H^*(\sigma_2)H(\sigma_1)\rangle = \exp\left[-\tfrac{1}{2}(\Delta\sigma T_{rms})^2\right], \qquad (8.7.8)$$

where $T_{rms}^2 = \langle T^2\rangle - \langle T\rangle^2$, and putting (8.7.8) into (8.7.2) yields

$$\langle I(\tau)\rangle = \exp\left[-\tfrac{1}{2}(\tau/T_{rms})^2\right]. \qquad (8.7.9)$$

Thus this simple example yields $I(\tau)$ with an extent of T_{rms} in τ, and an $F(\tau')$ with a zero spread in τ'. This is to be interpreted as the case where a δ-function is emitted from the source, and the receiver records a δ-function arrival (no *spread*) that *wanders* in τ by an amount T_{rms}, where T_{rms} is the rms variation in time of arrival of the pulse.

Note that in the geometrical-optics region $\sigma T_{rms} = \Phi$ so that we repeat (8.6.8):

$$\langle H^*(\sigma_2)H(\sigma_1)\rangle = \exp\left[-\tfrac{1}{2}(\Delta\sigma/\sigma)^2\Phi^2\right]. \qquad (8.7.10)$$

The general formula for $\langle I(\tau)\rangle$, (8.7.3), can now be interpreted as follows: the full time extent exhibited by $I(\tau)$ is made up of two parts; *wander* of the pulse corresponds to the exponential factor in the integrand, while *spread* of the pulse is a consequence of the Q factor. This interpretation is in keeping with the discussion in § 8.6 that showed Q to be a result of many-path interference, which is the only source of pulse spreading in the saturated region.

We can reinforce this interpretation with our second example. Suppose we are in the fully saturated region. Then the Q factor dominates the pulse extent and (8.7.3) can be done by approximate contour integration. The result, as discussed above, is that $\langle I(\tau)\rangle$ has an *extent* of $(\pi\eta_0)^{-1}$. Now consider the *spread*, as measured by $F(\tau')$. In the saturated region $H(\sigma)$ is an n-point Gaussian random

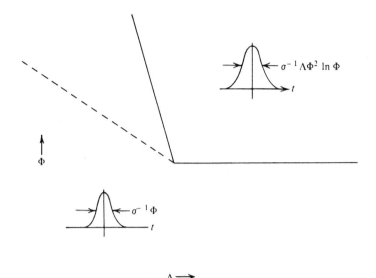

Fig. 8.14. Received-pulse extent for a δ-function transmission, in different regions of Λ–Φ space. This would be observed if many pulses were averaged together (a statistical ensemble).

variable (Chapter 9) so that the expectation value of the product of four H functions, as in (8.7.5), can be expressed in terms of a sum of products of two-point expectations. The result is (Chapter 12)

$$\langle F(\tau')\rangle = \delta(\tau') + A \int d\Delta\sigma |Q(\Delta\sigma/\eta_0)|^2 \exp{(-i\Delta\sigma\tau')}. \quad (8.7.11)$$

Thus the spread function has two contributions: (1) a replica of the emitted pulse; and (2) a spread term with characteristic width $(\pi\eta_0)^{-1}$. Any realizable experiment will not have a δ-function input pulse, but rather a narrow pulse with finite height N. The normalization is such that the spread term at $\tau' = 0$ gives N;

$$A \int d\Delta\sigma |Q(\Delta\sigma/\eta_0)|^2 = N. \quad (8.7.12)$$

We have shown that in the fully saturated region the spread has a value approximately equal to the extent. Therefore each realization of the ensemble has a pulse duration of the same amount as the pulse extent after averaging over the ensemble.

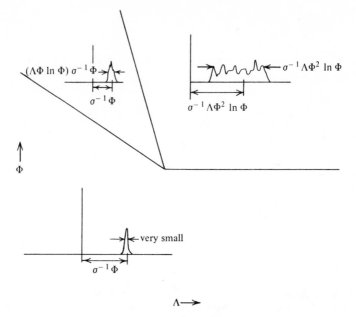

Fig. 8.15. Examples of individual received pulses in different regions of Λ–Φ space. Averaging many of these pulses yields the pulse extent shown in Fig. 8.14.

The extension of the above analysis for finite width input pulses is straightforward; also one may consider the cross-correlation between pulses transmitted at different times or received at different points in space (Dashen, 1977).

Fig. 8.14 shows examples of pulse transmission in the different regions of Λ–Φ space. An overall average of many pulses is shown, corresponding to *pulse extent.*

Examples of single pulse transmissions (Fig. 8.15) illustrate the separation between *spread* and *wander*.

Deterministic multipath spread

If more than one deterministic ray connects source and receiver, their travel times are not necessarily equal, and hence a transmitted pulse will be spread on arrival. An estimate of the spread due to this effect in the canonical sound channel can be made using the results of Chapter 4:

$$\Delta T = C^{-1} \Delta S = C^{-1} R \, \Delta(S_D/R_D), \qquad (8.7.13)$$

where Δ refers to the change between different deterministic rays. Using (4.1.5), (4.1.7) and (4.1.10) we have

$$\Delta T \approx (2\varepsilon/3)(R/C), \qquad (8.7.14)$$

for a SOFAR transmission with maximum $\phi_a^2 = 4$ and minimum $\phi_a^2 = 0$. As an example $\Delta T \approx 1$ s at $R = 400$ km, giving a coherent bandwidth of 1 Hz compared with 50 Hz from single-deterministic-path fluctuations (8.6.13).

Other variables

The above analysis of spread and wander in τ can be carried over exactly to spread and wander in the other conjugate variables θ_H, θ_V, and Doppler shift, γ. In a two-dimensional analysis of θ_H and θ_V these concepts are known as *speckle interferometry*.

CHAPTER 9

MULTIPATH EFFECTS AND n-POINT GAUSSIAN STATISTICS

In this chapter we develop the consequence of the received wavefunction being an n-point Gaussian random variable. The theory applies to cases where ψ is formed by the interference of many uncorrelated rays. Two physical effects can cause such a situation to arise: many *sporadic micropaths* because the sound transmission is in the saturated region; or many *deterministic* rays due to the sound channel in a long-range-propagation case. The rays must be uncorrelated in their fluctuations and each ray must have an overall phase fluctuation that is greater than one cycle. In the case of sporadic multipath these requirements are satisfied for full saturation, for spatial separations in partial saturation, and under certain conditions (see end of § 9.4) for time separations in partial saturation.

The statistics of ψ under the assumption of Gaussian behavior may be described entirely in terms of two-point functions (§ 9.1). Experimental comparisons may be made either by splitting ψ into real and imaginary parts (Cartesian statistics, § 9.2), or by using phase and amplitude (§ 9.3). However the statistics of phase and amplitude are more difficult to determine because of the sensitivity to *fadeout* effects. The precise definition of an n-point Gaussian random variable is given in § 9.4.

9.1. Statistics of a wavefunction obeying n-point Gaussian statistics

All vertical coordinates are taken to be close to some fixed point z_r and in what follows Φ, Λ, D, etc. are understood to be calculated for rays going from $(0, z_s)$ to (R, z_r). See Fig. 9.1.

The wavefunction ψ is a function of y, z, t and the frequency σ where the time dependence comes from the fluctuations. It is convenient to use a shorthand notation $\psi(1)$ to stand for

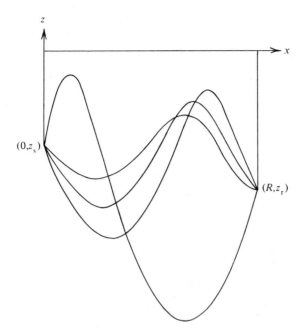

Fig. 9.1. A source–receiver geometry with many deterministic rays.

$\psi(y_1, z_1, t_1, \sigma_1)$. We have

$$\langle \psi \rangle = \langle \psi^* \rangle = 0,$$

$$\langle \psi(1)\psi(2) \rangle = \langle \psi^*(1)\psi^*(2) \rangle = 0, \qquad (9.1.1)$$

$$\langle \psi^*(1)\psi(2) \rangle \neq 0.$$

Higher moments in n-point Gaussian statistics

The average of a product of $n\psi$s and $m\psi^*$s vanishes unless $m = n$. The simplest nonvanishing higher-order correlation is then

$$\langle \psi^*(1)\psi(2)\psi^*(3)\psi(4) \rangle = \langle \psi^*(1)\psi(2) \rangle\langle \psi^*(3)\psi(4) \rangle$$

$$+ \langle \psi^*(1)\psi(4) \rangle\langle \psi^*(3)\psi(2) \rangle. \qquad (9.1.2)$$

The general nonvanishing correlation can be written as

$$\left\langle \prod_{j,k=1}^{n} \psi^*(j)\psi(k) \right\rangle.$$

Let $\bar{k}$ be a permutation of the indices k. For example if $n = 3$ and the permutation is $(1, 2, 3) \rightarrow (3, 1, 2)$ then $\bar{1} = 3$, $\bar{2} = 1$, and $\bar{3} = 2$ or if

the permutation is $(1, 2, 3) \rightarrow (2, 1, 3)$ then $\bar{1} = 2, \bar{2} = 1$, and $\bar{3} = 3$.
With this notation the correlation is

$$\left\langle \prod_{j,k=1}^{n} \psi^*(j)\psi(k) \right\rangle = \sum_{\text{perms}} \prod_{j,k=1}^{n} \langle \psi^*(j)\psi(\bar{k}) \rangle, \qquad (9.1.3)$$

where the sum is over all $n!$ possible permutations of the labels k. In
particular, for the intensity $I = \psi^*\psi$ at a given point

$$\langle (I)^n \rangle = n! (\langle I \rangle)^n. \qquad (9.1.4)$$

Calculation of two-point functions

The problem now reduces to the calculation of a two-point
function; the second-order correlation $\langle \psi^*(2)\psi(1) \rangle$. We use as an
example the case where points (1) and (2) are separated only in
time. To each ray connecting the source and receiver we can assign a
ψ_n and set $\psi = \sum_n \psi_n$. In terms of the ψ_n the second-order correlation is (see Chapter 8):

$$\langle \psi_n^*(t_1)\psi_m(t_2) \rangle = \delta_{mn}\langle I_n \rangle \exp\left[-\tfrac{1}{2}D(t_1 - t_2)\right] \qquad (9.1.5)$$

where D is understood to be computed for the appropriate ray.
   The approximation

$$D(t) \approx \nu^2 t^2 \qquad (9.1.6)$$

is always valid in the multipath situation, where $\nu = \dot{\Phi} = \Phi/\tau$ (see
(7.5.15)). That is, we may interpret ν as *the rate-of-change of phase
for a single ray*. Assuming either that there is a single unperturbed
ray with sporadic multipath or that ν^2 is roughly the same for all
deterministic rays we have, upon suppressing all variables except
time

$$\langle \psi^*(t_2)\psi(t_1) \rangle = \exp\left[-\tfrac{1}{2}\nu^2(t_2 - t_1)^2\right], \qquad (9.1.7)$$

where the result has been normalized to unity intensity. It is
disappointing that for typical conditions the multipath statistics
provide only such limited information about the ocean medium,
namely the parameter ν.

9.2. Cartesian statistics

Defining the Cartesian components

$$\psi(1) = X(1) + iY(1) \qquad (9.2.1)$$

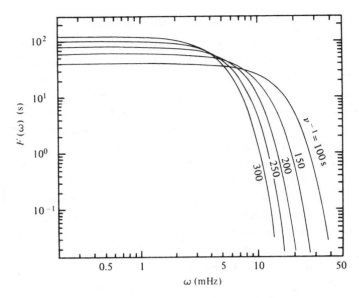

Fig. 9.2. Spectrum of a Cartesian component of the wavefunction for different values of ν.

we now use time separations as an example. From (9.1.7) it follows that

$$\langle X(t)X(t+\tau)\rangle = \tfrac{1}{2}\exp\left(-\tfrac{1}{2}\nu^2\tau^2\right). \tag{9.2.2}$$

The Fourier transform is the spectrum of $X(t)$, namely

$$F_X(\omega) = (2\pi)^{-\frac{1}{2}}\nu^{-1}\exp\left(-\tfrac{1}{2}\omega^2/\nu^2\right); \tag{9.2.3}$$

and similarly for $F_Y(\omega)$. The advantage of the Cartesian spectra over the more traditional polar coordinate representation involving intensity and phase (to be discussed later) is that the Cartesian spectra involve the single-path quantity ν in a simple way. Fig. 9.2 shows plots of (9.2.3) drawn for indicated values of ν^{-1}. An oceanographic model is needed to estimate expected values of ν^{-1}. Results for $\dot{\Phi}$ (which is identical to ν) from the internal-wave model are calculated in Chapter 7. Thus the Cartesian statistics are determined.

9.3. Intensity and phase statistics

The log-intensity, ι, and phase, ϕ, are defined by (8.1.6). We would like to determine the statistical behavior of ι and ϕ under the assumption of Gaussian statistics. The overall variance of ι is 5.6 dB as expressed in (8.1.9), but the phase ϕ executes a random walk and therefore grows without bound. This growth is associated with phase jumps near *fadeouts*, which result in phase spectra that are divergent at low frequency as the inverse-square power; the phase spectra are therefore experimentally difficult to measure. A more useful quantity is rate-of-change of phase, $\dot{\phi}$, which has well-behaved low-frequency behavior. Still, $\dot{\phi}$ will grow without bound due to the high-frequency behavior; however this divergence is only logarithmic and the finite upper limit inherent in any experimental measurement sets a natural upper cutoff to the spectrum. (See Dyson, Munk and Zetler, 1976.)

Under our basic assumptions the multipath components X and Y and their rates-of-change $\dot{X}$ and $\dot{Y}$ are independent Gaussian random variables. Using this information, Dyson, Munk and Zetler (1976) derive a correlation between the rates of change of ι and ϕ

$$C = \frac{\langle |\dot{\iota}|\, |\dot{\phi}| \rangle}{(\langle \dot{\iota}^2 \rangle \langle \dot{\phi}^2 \rangle)^{\frac{1}{2}}} = \frac{2}{\pi} = 0.63. \tag{9.3.1}$$

This relation is independent of the details of the fadeouts, but to obtain further information about the behavior of ι and ϕ we must examine the fadeouts more closely.

Fadeouts

The statistical behavior of ι and ϕ is dominated by the effects of *fadeouts*, which are brief periods during which both X and Y are small and ϕ is rapidly changing. It is convenient to define a fadeout precisely as a time-interval in which

$$A < \varepsilon \tag{9.3.2}$$

where ε is an arbitrarily chosen threshold fraction. (The intensity drop is $F = 20 \log_{10} \varepsilon^{-1}$ in dB, so $\varepsilon = 0.1$ corresponds to a 20 dB fadeout.) The model predicts the following statistical properties of fadeouts greater than 10 dB, based on the fact that, to a good approximation, the wavefunction ψ travels by the origin of the complex plane in a straight line (see § 9.4).

(i) The fraction of time occupied by fadeouts is

$$p(\varepsilon) = \varepsilon^2. \qquad (9.3.3)$$

(ii) The average duration of a fadeout is

$$\tau = \tfrac{1}{2}\pi^{\frac{3}{2}} \varepsilon \nu^{-1}. \qquad (9.3.4)$$

(iii) The average interval between fadeouts is

$$T = [\tau/p(\varepsilon)] = \tfrac{1}{2}\pi^{\frac{3}{2}}(\varepsilon\nu)^{-1}. \qquad (9.3.5)$$

Spectra

If ψ is an n-point Gaussian random function, then all the probability distributions are determined in terms of known two-point functions, in particular for the Cartesian components $X(t)$, $Y(t)$, $X(0)$, $Y(0)$. (We use time separation as an example.) From this multidimensional probability distribution the spectra of phase and amplitude may in principle be calculated, but in fact the calculation is difficult. One way of obtaining the answer is by numerical simulation of the physical basis for the Gaussian statistics, namely a large number of rays, each with an rms rate-of-change of phase, ν.

A numerical experiment on multipath interference has been performed, and the resulting spectra have been fitted to parametrized analytic forms (Dyson, Munk and Zetler, 1976).

Figs. 9.3 and 9.4 show the results of numerical experiments in multipath interference. The singlepath series $\delta\phi_i(t)$ were generated from random numbers for two cases: (i) a band limited white spectrum; and (ii) an ω^{-2} spectrum (by accumulating random $\delta^2\phi_i$). The singlepath phase series are formed by $\phi_i(t) = \sum_{t=0}^{t} \delta\phi_i(t)$, and the multipath according to

$$X(t) = \sum_{i=1}^{10} R_i \cos \phi_i, \qquad (9.3.6)$$

and similarly for $Y(t)$, with R_i arbitrarily set to 0.1. Spectra were computed for X, Y, $\dot{\phi}$, ι. This computation was repeated 10 times (using of course, different random noise series), and an average of the spectra so obtained has been plotted. The results are essentially the same for the ω^0 and ω^{-2} spectra of $\dot{\phi}_i$, as expected. (Internal waves would predict an ω^{-1} spectrum.)

For both cases we have taken $\nu^2 = \langle \dot{\phi}_i^2 \rangle = 1/(5 \text{ min})^2$. The input series consist of 2880 terms each, interpreted as a one day record at

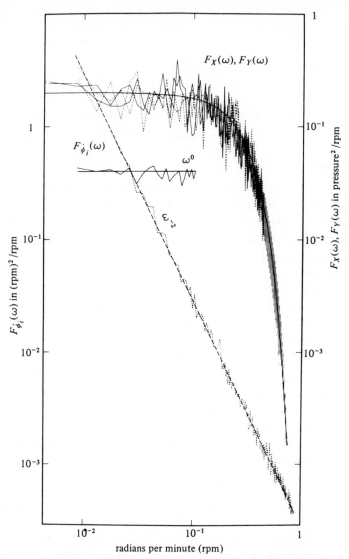

Fig. 9.3. Computer simulation of multipath statistics. The curves labeled
$F_{\dot{\phi}_i}(\omega)$ give the average spectra of the random single-path input functions
$\phi_i(t)$; solid: a band-limited (2 to 24 cpd) uniform spectrum, and dashed:
an ω^{-2} spectrum above 2 cpd. For both cases $\langle \dot{\phi}_i^2 \rangle = 1/(5 \text{ minutes})^2$. The
resulting spectra of the multipath Cartesian components $X(t)$ and $Y(t)$
(solid for ω^0, dashed for ω^{-2}) are in good accord with the predicted
Gaussian behavior. The shaded band gives limits of computed spectra at
high frequencies. (From Dyson, Munk and Zetler, 1976.)

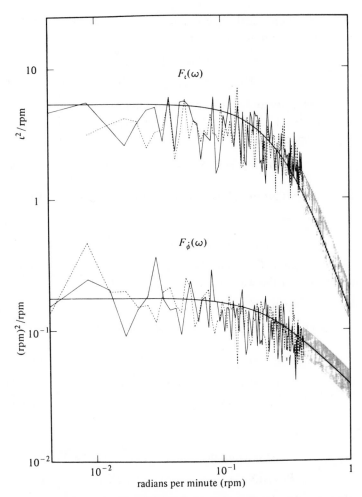

Fig. 9.4. Spectra of multipath intensity and rate-of-phase from the numerical experiment, corresponding to $F_{\phi_i}(\omega) \sim \omega^0$ (solid) and ω^{-2} (dashed), respectively. (From Dyson, Munk and Zetler, 1976.)

$\frac{1}{2}$-minute intervals. This sampling rate δt was chosen by trial and error to avoid ambiguities in multipath phase during occasional fadeouts. It would thus appear that $\nu \, \delta t = \frac{1}{10}$ is a good choice for a field experiment.

Fig. 9.3 shows the average of the 100 input spectra, and the associated Cartesian spectra according to (9.2.3); these provide a check on the numerical experiment. The spectra in Fig. 9.4 have been fitted by the following forms.

$$F_{\dot{\phi}}(\omega) = \nu^2(\omega^2 + 1.27\nu^2)^{-\frac{1}{2}} \tag{9.3.7}$$

$$F_\iota(\omega) = 4\nu^2(\omega^2 + 2.43\nu^2)^{-\frac{3}{2}}. \tag{9.3.8}$$

Thus the intensity and phase spectra are determined.

Adams (1976) has presented an exact series expression for the phase rate spectrum:

$$F_{\dot{\phi}}(\omega) = \tfrac{1}{2}\nu\pi^{-\frac{1}{2}} \sum_{n=1}^{\infty} n^{-\frac{3}{2}} \exp\left(-\omega^2/4n\nu^2\right) \tag{9.3.9}$$

and has shown that (9.3.7) is a good approximation to (9.3.9).

9.4. n-point Gaussian statistics

If ψ obeys n-point Gaussian statistics the probability of observing certain values of the wavefunction at n points is

$$P[\psi(1), \ldots, \psi(n)]\,\mathrm{d}V$$
$$= [\det(\pi M)]^{-1} \exp\left[-\sum_{i,j=1}^{n} \psi^*(i)M_{ij}^{-1}\psi(j)\right]\mathrm{d}V, \tag{9.4.1}$$

$$M_{ij} = \langle \psi^*(i)\psi(j)\rangle, \tag{9.4.2}$$

$$\mathrm{d}V = \mathrm{d}X_1\,\mathrm{d}Y_1\,\mathrm{d}X_2\,\mathrm{d}Y_2 \ldots \mathrm{d}X_n\,\mathrm{d}Y_n, \tag{9.4.3}$$

$$\psi(j) = X_j + \mathrm{i}Y_j. \tag{9.4.4}$$

One-point functions

In the case $n = 1$, M is a number equal to unity, and (9.4.1) becomes

$$P(\psi)\,\mathrm{d}X\,\mathrm{d}Y = \pi^{-1}\exp\left(-|\psi|^2\right)\mathrm{d}X\,\mathrm{d}Y \tag{9.4.5}$$

which shows directly that the phase of ψ is random uniformly over 2π and that the intensity has an exponential distribution, as in (8.3.7).

Two-point functions

In the case $n = 2$, M is a 2×2 matrix that can be inverted rather easily;

$$M = \begin{pmatrix} 1 & M_{12} \\ M_{12} & 1 \end{pmatrix}; \quad M^{-1} = (1 - M_{12}^2)^{-1}\begin{pmatrix} 1 & -M_{12} \\ -M_{12} & 1 \end{pmatrix}. \tag{9.4.6}$$

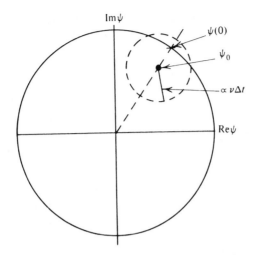

Fig. 9.5. Schematic of a Gaussian two-point function. At time zero the wavefunction is at $\psi(0)$. The dashed line indicates a line of constant probability of finding the wavefunction at time Δt.

It can then be easily shown that the probability of observing $\psi(2)$ given that $\psi(1)$ has been observed is

$$P(\psi_2, \psi_1)\, dX_2\, dY_2 = (\pi\sigma)^{-1} \exp\left(-|\psi_2 - \psi_0|^2/\sigma\right) dX_2\, dY_2$$
(9.4.7)

where

$$\sigma = 1 - M_{12}^2; \quad \psi_0 = M_{12}\psi(1); \quad M_{12} = \langle \psi^*(2)\psi(1)\rangle.$$
(9.4.8)

As an example consider two points separated in time so that $M_{12} = \exp\left[-\frac{1}{2}D(t)\right]$. Our result implies that the probability of observing a certain value for the wavefunction at time t is a two-dimensional Gaussian centered on ψ_0, which is the wavefunction at time zero multiplied by a factor that brings the center point closer to the origin (Fig. 9.5). The width of the Gaussian increases with time from zero at $t = 0$ to unity at $t = \infty$. It is important to remember that the conditions under which our Gaussian treatment is valid require that two points that are separated by large time intervals be statistically independent; that is $D(\infty)$ is a very large number. The distribution at very large times, when the two points are statistically independent, is identical to the one-point distribution (9.4.5); that is, a measurement taken a long time in the past has no effect on the probability distribution of the wavefunction now.

Three-point functions

In the case $n = 3$, M is a 3×3 matrix. The analogy continues, but we give here only one example that can be useful. Consider two measurements of ψ at $t = 0$ and $t = \varepsilon$. What is then the probability distribution of $\psi(t)$, given $\psi(0)$ and $\psi(\varepsilon)$? This could be calculated directly, but a simplification can be made by noticing that if ε is small, measuring $\psi(\varepsilon)$ is equivalent to a measurement of $\dot{\psi}$ at $t = 0$. That is, a measurement of ψ and $\dot{\psi}$ at $t = 0$ is equivalent to a two-point measurement; this choice of two points has the advantage that $\langle \dot{\psi}(0)\psi^*(0)\rangle = 0$, simplifying the matrix inversion. The result is

$$P(\psi_2, \psi_1, \dot{\psi}_1)\, dX_2\, dY_2 = (\pi\sigma')^{-1} \exp\left(-|\psi_2 - \psi_0'|^2/\sigma'\right) dX_2\, dY_2,$$

$$(9.4.9)$$

where

$$\sigma' = 1 - M_{21}^2 - \alpha M_{23}^2\,; \quad \psi_0' = M_{21}\psi(0) + \alpha M_{23}\dot{\psi}(0),$$

$$(9.4.10)$$

and

$$M_{21} = \langle \psi^*(t)\psi(0)\rangle\,; \quad M_{23} = \langle \psi^*(t)\dot{\psi}(0)\rangle\,; \quad \alpha = -\langle \psi_0^* \dot{\psi}_0\rangle^{-1}.$$

$$(9.4.11)$$

For small times we may use $D(t) \approx \nu^2 t^2$ so that

$$M_{21} = \exp\left(-\tfrac{1}{2}\nu^2 t^2\right)\,; \quad M_{23} \approx \nu^2 t \exp\left(-\tfrac{1}{2}\nu^2 t^2\right)\,; \quad \alpha = \nu^{-2},$$

$$(9.4.12)$$

and

$$\sigma' = \tfrac{1}{2}\nu^4 t^4\,; \quad \psi_0' = \psi(0) + t\dot{\psi}(0), \qquad (9.4.13)$$

so that the most probable value of ψ is on a straight line from $\psi(0)$ in the direction $\dot{\psi}(0)$, see Fig. 9.6. The deviation from a straight line will be small if

$$t \ll \langle(\dot{\psi})^2\rangle^{\frac{1}{2}}/\nu^2, \qquad (9.4.14)$$

which, to first approximation, requires $t \ll \nu^{-1}$.

Time separations in partial saturation

In the partially saturated region the wavefunction ψ is not a Gaussian variable for time separations; rather (8.4.7) applies with

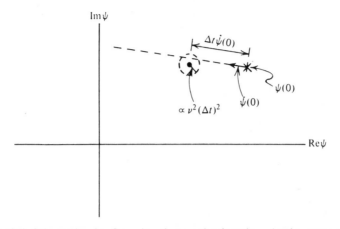

Fig. 9.6. Schematic of a Gaussian three-point function. At time zero the wavefunction is at $\psi(0)$ and its derivative points in the direction $\dot\psi(0)$. (This is the limit of two points very close in time.) The dashed line indicates a line of constant probability of finding the wavefunction at time Δt. The center of the circle is not on the dashed line because there is a tendency to be drawn toward the origin; see (9.4.10).

$\phi_0(t)$ being a Gaussian random variable with variances given by $\langle[\phi_0(t) - \phi_0(0)]^2\rangle = D(t)$. The functions $\chi(t)$ and $\phi_0(t)$ are statistically independent, but $\chi(t)$ is not in general a Gaussian random variable. Let us examine the moments of $\chi(t)$ in a manner similar to that used in § 9.1. We have that the average of a product of n χs and m χs vanishes unless $m = n$. The general nonvanishing correlation can be written as

$$\left\langle \prod_{k=1}^{n} \chi^*(t_k)\chi(t'_k) \right\rangle = \sum_{\text{perms}} K\left\{ t_0^{-1}\left[\sum_{k=1}^{n} (t_k - t'_{\bar{k}})^2 \right]^{\frac{1}{2}} \right\} \quad (9.4.15)$$

where $\bar{k}$ is defined in (9.1.3) and $K(t/t_0)$ is defined in § 8.4. Now we notice that if $K(t/t_0) = \exp\left(-\frac{1}{2}\nu'^2 t^2\right)$ then (9.4.15) will be identical with (9.1.3), and then the moments of χ will be identical with the moments of an n-point Gaussian variable. Therefore if $K(t/t_0)$ satisfies the above relation χ, and hence ψ, will satisfy n-point Gaussian statistics. For small time separations one can always approximate $K(t/t_0)$ in this manner, so that any statistical behavior at small time separations must be Gaussian. In particular, *fadeouts* happen over small time intervals and therefore fadeout distributions will be Gaussian in the partially saturated region.

PART IV

THEORY OF SOUND TRANSMISSION

The results summarized in Chapter 8 are presented in greater detail. The Rytov approximation is used in Chapter 10 to treat the unsaturated regions. The micromultipath theory is used in Chapters 11 and 12 to derive results for the saturated regions. Transport theory (Chapter 13) is a useful approximation for some applications, usually in the saturated regions. The transport equations, the method of moments, and the micromultipath theory are closely related mathematically. These relationships are described in Chapter 13.

CHAPTER 10

SUPEREIKONAL, OR RYTOV APPROXIMATION

We mentioned in Chapter 8 that an approximation scheme valid in the unsaturated region is the Rytov, or supereikonal method. Throughout the unsaturated region the scattering produces only one perturbed ray, so that the phase of the received signal is a useful quantity even where the acoustic scattering is strong enough to produce large phase fluctuations (but not yet so strong as to produce large intensity fluctuations). This chapter is devoted to a description of the supereikonal method, of how it is applied to the ocean, and of the results it yields for received phase and intensity fluctuations as a function of range and acoustic frequency. (The general result for complex amplitude correlations is discussed in Chapter 8.) This chapter is an expansion of Munk and Zachariasen (1976).

10.1. Isotropic ocean

In order to simplify the exposition of the method we begin by ignoring the deterministic sound channel. Thus we first discuss sound propagation through an infinite, isotropic, homogeneous ocean containing random fluctuations in sound speed. We therefore wish to solve the wave equation

$$(\nabla^2 + q^2)p(\mathbf{x}) = 2q^2\mu(\mathbf{x})p(\mathbf{x}) \equiv V(\mathbf{x})p(\mathbf{x}) \qquad (10.1.1)$$

subject to the boundary condition that as $\mathbf{x} \to 0$

$$p(\mathbf{x}) \to G(\mathbf{x}) = e^{iqx}/4\pi x \qquad (10.1.2)$$

which represents a unit strength point source at the origin.

Equation (10.1.1) may be cast into integral form through the use of the outgoing wave Green's function

$$(\nabla^2 + q^2)G(\mathbf{x} - \mathbf{y}) = \delta^3(\mathbf{x} - \mathbf{y}) \qquad (10.1.3)$$

resulting in

$$p(\mathbf{x}) = G(\mathbf{x}) + \int d^3y \ G(\mathbf{x} - \mathbf{y}) V(\mathbf{y}) p(\mathbf{y}). \qquad (10.1.4)$$

Iteration of this integral equation generates the perturbation series for $p(\mathbf{x})$, which is more conveniently written in Fourier transformed form as follows:

$$p(\mathbf{k}) = G(\mathbf{k}) + G(\mathbf{k}) \int (2\pi)^{-3} d^3k_1 \ V(\mathbf{k}_1) G(\mathbf{k} - \mathbf{k}_1)$$
$$+ G(\mathbf{k}) \int (2\pi)^{-3} d^3k_1 \int (2\pi)^{-3} d^3k_2 \ V(\mathbf{k}_1)$$
$$\cdot G(\mathbf{k} - \mathbf{k}_1) V(\mathbf{k}_2) G(\mathbf{k} - \mathbf{k}_1 - \mathbf{k}_2) + \dots \qquad (10.1.5)$$

where

$$p(\mathbf{k}) = \int d^3x \exp(-i\mathbf{k} \cdot \mathbf{x}) p(\mathbf{x}),$$
$$G(\mathbf{k}) = (k^2 - q^2 + i\varepsilon)^{-1}. \qquad (10.1.6)$$

The supereikonal approximation now consists of neglecting all momentum transfer correlations in the perturbation series. That is, we approximate $(\mathbf{k} - \mathbf{k}_1 - \mathbf{k}_2 - \dots - \mathbf{k}_n)^2 - q^2 + i\varepsilon$ by $k^2 - 2\mathbf{k} \cdot (\mathbf{k}_1 + \mathbf{k}_2 + \dots + \mathbf{k}_n) + k_1^2 + k_2^2 + \dots + k_n^2 - q^2 + i\varepsilon$, and neglect all terms of the form $\mathbf{k}_i \cdot \mathbf{k}_j$ when $i \neq j$. Note that the first approximation occurs in the second-order term in V. Once this simplification is made, the perturbation series can be summed exactly, and one obtains the result

$$p(\mathbf{x}) = (-i\pi)^{\frac{1}{2}} (8\pi^2)^{-1} \int_0^\infty \beta^{-\frac{3}{2}} d\beta$$
$$\cdot \exp \{i[\beta q^2 + (x^2/4\beta) + \beta I(\beta, \mathbf{x}) + i\varepsilon]\}, \qquad (10.1.7)$$

where

$$I(\beta, x) = \int (2\pi)^{-3} d^3k \ V(\mathbf{k}) \int_0^1 ds \exp[+i(s\mathbf{k} \cdot \mathbf{x} + \beta s(1-s)k^2)]. \qquad (10.1.8)$$

This expression constitutes the supereikonal approximation to the pressure. It is valid in the unsaturated region. An approximation to the range of validity is given by:

$$qR \gg 1, \quad qL \gg 1, \quad R < L/\mu_{\mathrm{rms}}. \qquad (10.1.9)$$

It is worth noting that if in (10.1.8) the $\beta s(1-s)k^2$ term is omitted from the exponent, we obtain

$$p(\mathbf{x}) = (4\pi x)^{-1} \exp\left[ix\left(q^2 + \int_0^1 V(s\mathbf{x}) ds\right)^{\frac{1}{2}}\right], \qquad (10.1.10)$$

which is the conventional WKB, or eikonal, or geometrical-optics approximation to the pressure. The primary virtue of the supereikonal form, therefore, is that it contains as limiting cases *both* the conventional eikonal *and* the complete first-order perturbation theory approximations. While (10.1.7) and (10.1.8) do constitute a closed-form solution for the pressure, the expressions are still a bit unwieldy, and further simplification is useful. To this end we may evaluate the integral in (10.1.7) by stationary phase (MZ76) to obtain

$$p(\mathbf{x}) = G(\mathbf{x}) \exp[X(\mathbf{x})], \qquad (10.1.11)$$

where

$$X(\mathbf{x}) = G^{-1}(\mathbf{x}) \int d^3\mathbf{x} \, G(\mathbf{x} - \mathbf{x}')V(\mathbf{x}')G(\mathbf{x}'). \qquad (10.1.12)$$

This is known as Rytov's approximation to the pressure. A direct derivation of it may be made by replacing the wave equation (10.1.1) by an equation for $\log[p(\mathbf{x})/G(\mathbf{x})]$ and solving this to first order in V. However, the justification for the approximation is somewhat obscure in this direct derivation; the supereikonal technique allows the neglected terms to be more clearly visualized.

In any event, depending on the appropriateness of the stationary-phase method one may use either the supereikonal expression (10.1.7) or the Rytov expression (10.1.11) to proceed further. We shall use (10.1.11) and (10.1.12). We are, of course, not directly interested in the pressure itself, but rather in the pressure averaged over many transmissions. Assuming that the fluctuations $\mu(\mathbf{x})$ obey Gaussian statistics, the averages of interest can be obtained directly from (10.1.11). In particular, the average pressure is

$$\langle p(\mathbf{x}) \rangle = G(\mathbf{x}) \exp[\tfrac{1}{2}\langle X^2 \rangle]; \qquad (10.1.13)$$

the mean-square pressure is

$$\langle p^2(\mathbf{x}) \rangle = G^2(\mathbf{x}) \exp[2\langle X^2 \rangle]; \qquad (10.1.14)$$

the absolute mean-square pressure is

$$\langle |p(\mathbf{x})|^2 \rangle = |G(x)|^2 \exp[2\langle (\operatorname{Re} X)^2 \rangle]. \qquad (10.1.15)$$

The mean-square phase fluctuations are

$$\langle \phi^2 \rangle = \langle (\operatorname{Im} X)^2 \rangle = \tfrac{1}{2}[\langle |X|^2 \rangle - \operatorname{Re} \langle X^2 \rangle], \qquad (10.1.16)$$

and the mean-square log-intensity fluctuations are

$$\langle \iota^2 \rangle - \langle \iota \rangle^2 = 4\langle (\operatorname{Re} X)^2 \rangle = 2[\langle |X|^2 \rangle + \operatorname{Re} \langle X^2 \rangle]. \quad (10.1.17)$$

Thus all the quantities in which we have an interest may be expressed in terms of the averages $\langle |X|^2 \rangle$ and $\langle X^2 \rangle$; or equivalently in terms of the averages $\langle X_1^2 \rangle$, $\langle X_2^2 \rangle$, and $\langle X_1 X_2 \rangle$ where $X = X_1 + iX_2$. We shall therefore direct our discussion toward obtaining expressions for these.

The Green's function G is given by (10.1.2); hence we may write for the quantity X the expression

$$X(\mathbf{x}) = (4\pi)^{-1} \int d^3 y \, xy^{-1} |\mathbf{x} - \mathbf{y}|^{-1} \exp [iq(y + |\mathbf{x} - \mathbf{y}| - x)] V(\mathbf{y}). \quad (10.1.18)$$

It is straightforward to evaluate $\langle |X|^2 \rangle$ and $\langle X^2 \rangle$ from this formula. The details are given in MZ76, and the results are

$$\langle |X|^2 \rangle = 2\pi q^2 R \int d^2 \mathbf{k}_T \, F(0, \mathbf{k}_T) = q^2 R \int_{-\infty}^{\infty} \rho(x) \, dx \quad (10.1.19)$$

$$\langle X^2 \rangle = -2\pi q^2 \int d^2 \mathbf{k}_T \, F(0, \mathbf{k}_T) \cdot \int_0^R dx \, \exp [i(k_T^2/q)(R - x)x/R)] \quad (10.1.20)$$

where $F(\mathbf{k})$ is the spectrum of sound-speed fluctuations (6.1.9).

The vector $\mathbf{k}_T$ is the component of $\mathbf{k}$ perpendicular to the straight line joining the origin to the point $\mathbf{x}$. (This is the unperturbed ray path in the flat ocean case.) The notation $F(0, \mathbf{k}_T)$ indicates that F is to be evaluated with the parallel component of $\mathbf{k}$ set equal to zero. Equations (10.1.19) and (10.1.20) constitute the central results of this section. They express the quantities of interest as integrals along unperturbed ray paths (in this case straight lines) of the Fourier transform of the correlation function $F(\mathbf{k})$ multiplied by rather simple geometrical factors. As we shall see later, entirely parallel expressions obtain in the more difficult case of an inhomogeneous background medium.

Regions of validity

The expression for $\langle |X(\mathbf{x})|^2 \rangle$, (10.1.19), is precisely the same result for this quantity obtained by using geometrical optics to compute

$X(\mathbf{x})$ itself, and then calculating $\langle|X(\mathbf{x})|^2\rangle$ from this. It is thus identical to the conventional geometric-optics expression for the mean-square phase fluctuations, which we called Φ^2 in Chapter 6 (6.2.5):

$$\langle|X|^2\rangle \equiv \Phi^2. \tag{10.1.21}$$

In contrast, (10.1.20) is *not* what one obtains for $\langle X(\mathbf{x})^2\rangle$ from geometrical optics. Geometrical optics for this quantity is recovered if one expands the exponential in (10.1.20), a procedure that evidently is valid only if $(k_T^2/q)x(R-x)/R \ll 1$. Since $k_T \sim 1/L_V$ and $x(R-x) \sim R^2$ this condition can be more familiarly written as $\lambda R \ll L_V^2$, which we recognize as the Fresnel condition under which the geometrical-optics approximation for $X(\mathbf{x})$ itself is valid.

Note that the quantity $R/x(R-x)$ is the phase curvature A for the flat ocean. Thus

$$\langle|X|^2\rangle + \langle X^2\rangle = 2\pi q^2 \int_0^R dx \int d^2\mathbf{k}_T \, F(0, \mathbf{k}_T) \cdot [1 - \exp(ik_T^2/qA)]$$

$$\tag{10.1.22}$$

and in view of (10.1.17)

$$\langle\iota^2\rangle - \langle\iota\rangle^2 = 4\pi q^2 \int_0^R dx \int d^2\mathbf{k}_T \, F(0, \mathbf{k}_T)(1 - \cos k_T^2/qA).$$

$$\tag{10.1.23}$$

Recalling our definition in Chapter 7 of the quantity $\Lambda \sim (qL_V^2 A)^{-1}$ and noting again the relation $k_T \sim 1/L_V$, we see that in the region $\Lambda \ll 1$, we have $\langle|X|^2\rangle \approx -\langle X^2\rangle$, which is the requirement that geometrical optics applies. In this regime intensity fluctuations are evidently small.

If the spectrum $F(0, \mathbf{k}_T)$ falls off sufficiently rapidly with k_T (for example exponentially) we may expand $\cos k_T^2/qA$ in (10.1.23). This yields the estimate $\langle\iota^2\rangle - \langle\iota\rangle^2 \sim \Phi^2\Lambda^2$. If, on the other hand, the spectrum is not so sharply cut off, but instead falls like a power of k_T, the cosine in (10.1.23) cannot be expanded. The behavior of the integral can, however, be investigated for power-law spectra $F(0, \mathbf{k}_T) \propto k_T^{-p}$, $1 < p < 3$, with the result that

$$\langle\iota^2\rangle - \langle\iota\rangle^2 \sim \Phi^2\Lambda^{p-1}. \tag{10.1.24}$$

The internal-wave spectrum corresponds to $p \approx 2$, so that intensity fluctuations, as calculated by the Rytov formula, become of order unity when $\Phi^2 \Lambda \sim 1$. We remind the reader that $\Phi^2 \Lambda \approx 1$ is the boundary between the saturated and unsaturated regions when $\Lambda < 1$, as defined in Chapter 8 (Fig. 8.5) from the micromultipath point of view.

Thus (10.1.20) constitutes an improvement over geometrical optics, while (10.1.19) coincides with geometrical optics. Conversely, geometrical optics for $\langle |X|^2 \rangle$ is valid out to a very large range, while geometrical optics for $\langle X^2 \rangle$ is valid only within the range $R \ll qL_V^2$, or, in other words, $\Lambda \ll 1$.

It is of interest to study (10.1.20) in the limit of very long range. As $R \to \infty$, the integral over dx can be approximately evaluated, and we find

$$\langle X^2(\mathbf{x}) \rangle \approx [iq^2 p(0)/2\pi](\gamma + \log 4qR - \tfrac{1}{2}i\pi) \sim i \log R,$$
(10.1.25)

where $\gamma = 0.577 \ldots$ is Euler's constant, while for small R, satisfying the Fresnel condition, we have the geometric-optics limit

$$\langle X^2(\mathbf{x}) \rangle \approx -2\pi q^2 \int d^2 k_{\mathrm{T}} F(0, \mathbf{k}_{\mathrm{T}})[R - (ik_{\mathrm{T}}^2/6q)R^2 + \ldots].$$
(10.1.26)

Between these limits (10.1.20) provides a smooth transition. In contrast to (10.1.25) we have from (10.1.19) the result

$$\langle |X|^2 \rangle \sim R$$

for both large and small x.

We end this section by emphasizing that the key result of this chapter on the unsaturated region is the evaluation of $\langle X^2 \rangle$, which cannot be calculated from knowledge of Λ, Φ and the phase-structure function D. Since $\langle |X|^2 \rangle$ is in fact equal to Φ^2, all our results for Φ^2 from Chapter 7 are applicable to $\langle |X|^2 \rangle$ in this chapter. Our results for $\langle X^2 \rangle$ on the other hand are restricted in validity to the unsaturated regime.

10.2. Anisotropic ocean

Now let us turn to the effects of the sound channel. That is, we must replace the non-fluctuating sound speed C in the isotropic case by a specified function of position $C(\mathbf{x})$.

The wave equation for the pressure, which is our starting point, now becomes altered from (10.1.1) to the equation

$$(\nabla^2 + q^2(\mathbf{x}))p(\mathbf{x}) = V(\mathbf{x})p(\mathbf{x}), \tag{10.2.1}$$

though still with the same boundary condition.

We must first study the non-fluctuating part of the problem, to evaluate the Green's function in the presence of the sound channel. This satisfies

$$[\nabla^2 + q^2(\mathbf{x})]G(\mathbf{x}, \mathbf{y}) = \delta^3(\mathbf{x} - \mathbf{y}); \tag{10.2.2}$$

note that G is no longer a function only of $\mathbf{x} - \mathbf{y}$ as it was in the homogeneous case.

We shall assume that geometrical optics provides a good approximation to the non-fluctuating sound channel problem. This means that we can represent $G(\mathbf{x}, \mathbf{y})$ as a sum of contributions from each ray joining $\mathbf{x}$ and $\mathbf{y}$. To be specific, we may write

$$G(\mathbf{x}, \mathbf{y}) = \sum_{i=1}^{n(\mathbf{x}, \mathbf{y})} G_i(\mathbf{x}, \mathbf{y}), \tag{10.2.3}$$

where $n(\mathbf{x}, \mathbf{y})$ is the number of rays and G_i is the contribution of the ith ray. We have, in particular, for rays joining the origin and $\mathbf{x}$,

$$G_i(\mathbf{x}, 0) = K_i(\mathbf{x}, 0) \exp i \int_0^{\mathbf{x}} ds \, q[\mathbf{x}_i(s)], \quad i = 1 \ldots n(\mathbf{x}), \tag{10.2.4}$$

where ds is an element of path length along the ray, $\mathbf{x}_i(s)$ is the ith ray joining 0 to $\mathbf{x}$, and K_i is a normalization factor.

Now when the fluctuations are turned on, the signals traveling on each of the rays joining the origin to the point of observation $\mathbf{x}$ are subject to small-angle scatterings by the perturbing potential $V(\mathbf{x})$. The signals are thus deflected slightly from the undisturbed rays by each interaction with V. The repeated action of V therefore produces, on each ray, a sort of random walk of the signal away from the original ray. When we average over an ensemble of perturbations V, the disturbed signals will fill up a tube surrounding the undisturbed ray. Provided that these tubes around each of the original rays do not overlap, the received pressure will be a sum of contributions from each ray tube. (Such tubes exist, of course, in the homogeneous case as well, but there they never overlap.)

We may estimate the radius of a ray tube as follows. The mean free path d between interactions of the signal traveling along a given ray with the perturbing potential V is of the order of $(\mu q)^{-1}$. Hence over a range R the number of scatterings is $n = R/d$. The average deflection angle due to each scattering is of order $1/qL_V$ vertically and $1/qL_H$ horizontally, where L_V and L_H are the vertical and horizontal correlation lengths of the sound-speed fluctuations. Since the process is a random walk, the net displacement due to n collisions (when n is large) is proportional to $\sqrt{n}$, and hence the vertical and horizontal extents of the tube are, roughly,

$$r_V \sim (R/q\mu)^{\frac{1}{2}}(qL_V)^{-1}, \quad r_H \sim r_V(L_V/L_H).$$

Let us assume that the vertical extent of the tubes is small enough so that the tubes remain distinct. Then the pressure at $\mathbf{x}$ is the sum of contributions from each tube;

$$p(\mathbf{x}) = \sum_{i=1}^{n(x)} p_i(\mathbf{x}), \tag{10.2.5}$$

where $n(\mathbf{x})$ is the number of unperturbed rays joining the source to the point $\mathbf{x}$. We shall be interested in $p_i(\mathbf{x})$.

Munk and Zachariasen (1976) derive the analog of (10.1.11) and (10.1.12) by introducing an unperturbed Green's function $G_i(\mathbf{x}, \mathbf{y})$ which represents the field at $\mathbf{x}$ due to a source at $\mathbf{y}$ that emits energy only in a small solid angle around the direction of the ith unperturbed ray. The result is

$$P_i(\mathbf{x}) = G_i(\mathbf{x}, 0) \exp\left[G_i^{-1}(\mathbf{x}, 0) \int_{i\text{th ray tube}} \right.$$
$$\left. \cdot \, d^3\mathbf{y} \, G(\mathbf{x}, \mathbf{y})V(\mathbf{y})G(\mathbf{y}, 0)\right]. \tag{10.2.6}$$

This is evidently the generalization of the Rytov formula to the situation of an inhomogeneous background and many rays. The expression clearly fails if the range is so large that the ray tubes overlap; otherwise the validity conditions are the same as those in the homogeneous background case.

We shall now use (10.2.6) to calculate the various averages of interest for the contribution of a single ray tube to the pressure in the presence of the sound channel. We shall, for simplicity, drop the index i, though we should keep in mind that when there are several unperturbed ray paths their contributions are to be added to obtain the total pressure. Our interest, then, will be in the statistical fluctuations of the contributions of a single ray, or rather a single ray tube.

As in the isotropic case, we define

$$X(\mathbf{x}) = G^{-1}(\mathbf{x}, 0) \int d^3\mathbf{y}\, G(\mathbf{x}, \mathbf{y}) V(\mathbf{y}) G(\mathbf{y}, 0), \qquad (10.2.7)$$

and we wish to compute $\langle X^2 \rangle$ and $\langle |X|^2 \rangle$. We recall that assuming geometrical optics to be a valid approximation for the non-fluctuating background permits us to write the Green's function as

$$K(\mathbf{x}, \mathbf{y}) \exp[iqS(\mathbf{x}, \mathbf{y})], \qquad (10.2.8)$$

where

$$qS(\mathbf{x}, \mathbf{y}) = \int_{\mathbf{x}}^{\mathbf{y}} ds\, q[\mathbf{x}(s)], \qquad (10.2.9)$$

and where the normalization factor is given to a sufficient approximation (MZ76) by

$$K(\mathbf{x}, \mathbf{y}) = (4\pi|\mathbf{x} - \mathbf{y}|)^{-1}. \qquad (10.2.10)$$

In (10.2.9) the line integral is along the ray of interest joining the points $\mathbf{x}$ and $\mathbf{y}$.

The determination of $\langle X^2 \rangle$ and $\langle |X|^2 \rangle$ follows the same method as for the homogeneous case (again see MZ76). We find

$$\langle |X(\mathbf{x})|^2 \rangle = 2\pi q^2 \int_0^{\mathbf{x}} ds \int d^2\mathbf{k}_T(s) F[\mathbf{k}_T(s), \mathbf{Y}(s)] = \Phi^2,$$

$$(10.2.11)$$

in complete parallel to the homogeneous case. Here the line integral on ds is along the unperturbed ray, $\mathbf{k}_T(s)$ refers to the component of $\mathbf{k}$ perpendicular to the ray at s and $\mathbf{Y}(s)$ is a point on the ray at s. We also find

$$\langle X(\mathbf{x})^2 \rangle = -2\pi q^2 \int_0^{\mathbf{x}} ds \int d^2\mathbf{k}_T(s) F[\mathbf{k}_T(s), \mathbf{Y}(s)]$$

$$\cdot \exp[(i/q)k_{T_i}(s)k_{T_j}(s)A^{-1}(\mathbf{Y}(s))_{ij}]. \qquad (10.2.12)$$

The notation is as in (10.2.11), and the result is again in complete analogy to the homogeneous case. The matrix A_{ij} was defined in Chapter 7.

Most of the comments we made in § 10.1 concerning the results in the homogeneous background apply here as well. The expression for $\langle |X|^2 \rangle$ is again just that obtained in the geometrical-optics approximation, so that again $\langle |X|^2 \rangle = \Phi^2$, but that for $\langle X^2 \rangle$ is not. Geometrical optics for $\langle X^2 \rangle$ is valid provided that

$$(1/q)k_{T_i}(s)k_{T_j}(s)A^{-1}[\mathbf{Y}(s)]_{ij} \ll 1; \qquad (10.2.13)$$

this is the analog of the Fresnel condition. It is also the condition $\Lambda \ll 1$, as in the flat ocean case, and again, for $\Lambda \ll 1$ the condition $\Phi^2 \Lambda \sim 1$ (the boundary of the geometrical-optics region) corresponds to intensity fluctuations of order unity.

10.3. Channeled ocean

Let us apply our general results, (10.2.11) and (10.2.12), to the specific case of a channeled ocean with its associated cylindrical symmetry. We choose the z axis positive upward, and the unperturbed ray lies in the x, z plane. We confine ourselves to situations in which the source and receiver lie at the same depth. Thus the source is at the point $(x, y, z) = (0, 0, 0)$ and the receiver is at the point $(R, 0, 0)$ where R is the range. The unperturbed ray path joining these two points will be denoted $z(x)$; thus $z(0) = 0$, $z(R) = 0$ and $\theta(x)$ is the angle the ray makes with the horizontal at the point x. The element of path length along the ray is then given to sufficient accuracy by $ds = dx$ because we are dealing with small-angle rays (in keeping with the parabolic approximation). (We note that MZ76 used the exact relation $ds = \sec \theta \, dx$.)

The expressions for $\langle |X|^2 \rangle$ and $\langle X^2 \rangle$ both involve integrals of the Fourier components of the correlation function over a plane perpendicular to the ray path at each point along the ray. With our geometry, the wavenumber $\mathbf{k}(s)$ perpendicular to the ray has x, y and z components

$$\mathbf{k}_T(s) = (-k_z \tan \theta, k_y, k_z)$$

and the element of surface area perpendicular to the ray is

$$d^2\mathbf{k}_T(s) = dk_y \, dk_z \sec \theta \approx dk_y \, dk_z.$$

Thus (10.2.11) becomes

$$\langle |X|^2 \rangle = 2\pi q^2 \int_0^R dx \int_{-\infty}^{\infty} dk_y \int_{-\infty}^{\infty} dk_z$$
$$\cdot F[(-k_z \tan \theta(x), k_y, k_z), z(x)] = \Phi^2. \qquad (10.3.1)$$

We may express $\langle X^2 \rangle$ in a similar manner, except that the phase curvature, A, defined in Chapter 7, is required. The result is expressed in (10.2.12).

10.4. Internal-wave dominance

If we commit ourselves to the specific ocean model described in Chapter 3, it turns out to be more convenient to express the correlation function F in terms of the variables ω and j, the frequency and mode number of the internal waves, rather than the wavenumbers k_y and k_z. The transformation to these new coordinates is accomplished as follows: horizontal and vertical components of wavenumber

$$k_H = (k_z^2 \tan^2 \theta + k_y^2)^{\frac{1}{2}}, \quad k_V = k_z$$

have the approximate dispersion relations (see Chapter 3)

$$k_H = j\pi B^{-1}(\omega^2 - \omega_i^2)^{\frac{1}{2}}/n_0, \quad k_V = j\pi B^{-1} n/n_0.$$

Setting $\omega_L^2 = \omega_i^2 + n^2 \tan^2 \theta$, we also have the relation $k_y = (\pi j / n_0 B)(\omega^2 - \omega_L^2)^{\frac{1}{2}}$.

Then we note the definition

$$\int d^3\mathbf{k}\, F(k, z) = 2\pi \int_0^\infty k_y\, dk_y \int_{-\infty}^\infty dk_z\, F(0, k_y, k_z; z)$$

$$= \sum_j \int_{\omega_i}^n d\omega\, F(\omega, j; z) \qquad (10.4.1)$$

where $F(\omega, j; z)$ is the spectrum of sound-speed fluctuations (§ 3.6).

Thus (10.3.1) may be replaced by

$$\langle |X|^2 \rangle = 2q^2 \int_0^R dx \sum_j \int_{\omega_L}^n d\omega\, k_y^{-1} F(\omega, j; z). \qquad (10.4.2)$$

In this expression z, n, and θ are, of course, functions of x, to be evaluated at each point along the unperturbed ray from source to receiver. The lower limit of the frequency integral is ω_L because frequencies below that value cannot satisfy the dispersion relation with a real k_y if $k_x = k_z \tan \theta$. The factor of two arises from the equal contributions from positive and negative k_y.

A similar transformation may be carried out for $\langle X^2 \rangle$, the other quantity of interest. The only complication here is that it is necessary to use the matrix A_{ij}, introduced in Chapter 7, at each point along the ray. This matrix, in turn, is expressible in terms of the phase curvature A. Indeed, we can write the exponent in the integrand of (10.2.12) as

$$q^{-1}k_{T_i}(s)k_{T_j}(s)A^{-1}[\mathbf{Y}(s)]_{ij} = q^{-1}\{k_y^2[x(R-x)/R] + k_z^2 A^{-1}\}.$$

Hence, using the dispersion relations for horizontal and vertical components of wavenumbers, we find

$$\langle X^2 \rangle = -2q^2 \int_0^R dx \sum_j \int_{\omega_L}^n d\omega \, k_y^{-1} F(\omega, j; z) \exp\{iq^{-1}(j\pi/Bn_0)^2$$
$$\cdot [(x(R-x)/R)(\omega^2 - \omega_L^2) + n^2 A^{-1}]\}. \tag{10.4.3}$$

Geometrical optics is valid when the exponent in this equation is much less than one; this is the analogue of the Fresnel condition in the channeled ocean.

Except when the receiver is in the vicinity of a caustic, it is in general the case that the k_y^2 term in the exponent, which is associated with horizontal spreading, is much smaller than the term with A^{-1}. This is because, as we shall see below, the spectrum $F(\omega, j; z)$ tends to weigh small values of ω much more heavily. Thus for most purposes we can ignore this term, and replace (10.4.3) by the much simpler expression

$$\langle X^2 \rangle = -2q^2 \int_0^R dx \sum_j 1/j \exp(i\beta j^2) \int_{\omega_L}^n d\omega \, k_y^{-1} F(\omega, j; z). \tag{10.4.4}$$

where

$$\beta = (\pi n/n_0 B)^2 (qA)^{-1}. \tag{10.4.5}$$

Equations (10.4.2) and (10.4.4) are as far as we can go toward computing the quantities of interest without committing ourselves to a particular spectrum F, and a particular sound channel and associated ray paths. Substitution of the spectrum of (3.6.5) into (10.4.2) yields (MZ76)

$$\langle |X|^2 \rangle = \langle \mu_0^2 \rangle \langle j^{-1} \rangle q^2 B R_0 F_1(R), \tag{10.4.6}$$

where

$$F_1(R) = 4(\pi^2 \omega_i n_0^2 R_0)^{-1} \int_0^R dx\, n^3 f_1(\Delta), \qquad (10.4.7)$$

$$f_1(\Delta) = (\Delta^2 + 1)^{-1} + \tfrac{1}{2}\Delta^2(\Delta^2 + 1)^{-\frac{3}{2}} \ln\{[(\Delta^2 + 1)^{\frac{1}{2}} + 1]/[(\Delta^2 + 1)^{\frac{1}{2}} - 1]\}, \qquad (10.4.8)$$

$$\Delta = (n/\omega_i)\tan\theta. \qquad (10.4.9)$$

$F_1(R)$ as defined here is a dimensionless number of order one when R is of the order of R_0, the range of a loop, $(R_0 = \tfrac{1}{2}B\pi\varepsilon^{-\frac{1}{2}} = 20.8$ km, see Chapter 4). It is for this reason that the factor R_0 has been explicitly separated out in (10.4.6).

The quantity $\langle j^{-1} \rangle$ represents the average of j^{-1} weighted by the internal-wave spectrum $H(j)$. We have

$$\langle j^{-1} \rangle = \sum_j j^{-1} H(j) = 0.730, 0.647, 0.519, 0.435, 0.379, 0.340 \qquad (10.4.10)$$

for $j_* = 0, 1, \ldots, 5$. An approximate expression is $\log(4j_*^2 + 1)/(\pi j_* - 1)$.

For axial rays, $\theta = \Delta = 0$, and $f_1(0) = 1$ and $n = n_1$ is a constant. Hence $F_1(R)$ becomes simply proportional to R; we have

$$F_1(R) = (4/\pi^2)(n_0/\omega_i)(n_1/n_0)^3(R/R_0) = 1.436\,(R/R_0) \equiv \bar{F}_1(R). \qquad (10.4.11)$$

For upward rays turning near the surface, the major contribution to the integral defining $F_1(R)$ comes from the ray apex (Fig. 10.1); here the equation of the ray is, approximately, $z(x) \approx z_a - (2r)^{-1}(x - x_a)^2$, and $\Delta = n(x - x_a)/\omega_i r$, where r is the ray curvature at the apex. The function $f_1(\Delta)$ cuts off rapidly with increasing Δ, and therefore we can write, approximately,

$$F_1(R) \approx \hat{F}_1(R) = 2(r/R_0)(n_a/n_0)^2 N \qquad (10.4.12)$$

where N is the number of upward loops.

For a single complete upward or downward loop with range R^+ or R^- respectively we denote $F_1(R^+)$ and $F_1(R^-)$ by F_1^+ and F_1^-, respectively. Then for a complete double loop, with range R_D, we have

$$F_1^{+-} \equiv F_1(R_D) = F_1^+ + F_1^-.$$

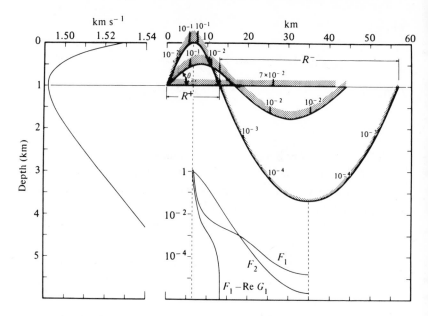

Fig. 10.1. The canonical sound channel (left) and the corresponding rays for $\theta = 12.7°$ (surface limited), $5.2°$ and $0°$ (axial ray). The contribution to F_1 from various parts along the three ray paths is indicated by the vertical extent of the shaded band (plotted logarithmically). F_1 is plotted separately at the bottom of the figure for the surface-limited ray, together with F_2 and $F_1 - \text{Re } G_1$, thus indicating the relative apex contributions toward mean-square phase, rate-of-phase change, and intensity. ($F_1 - \text{Re } G_1$ applies only to the source at $x = 0$ of a receiver at R^+.) (From Munk and Zachariasen, 1976.)

If the double loop has an apex near the surface, then $F_1^+ \approx \hat{F}_1$ and $F_1^- \approx 0$; thus

$$F_1^{+-} \approx \hat{F}_1.$$

The results of a numerical evaluation of F_1^+ and F_1^- as a function of ray angle θ_1 are shown in Fig. 10.2, as is the apex approximation $\hat{F}_1$, which is seen to be an excellent approximation for $\theta_1 \gtrsim 5°$. The largest value of F_1^+ occurs at a ray angle θ_1 near $2°$, and a corresponding apex depth of 750 m; deeper rays are reduced by the smaller value of n^3, rays of shallower apex are reduced by a smaller radius of curvature r.

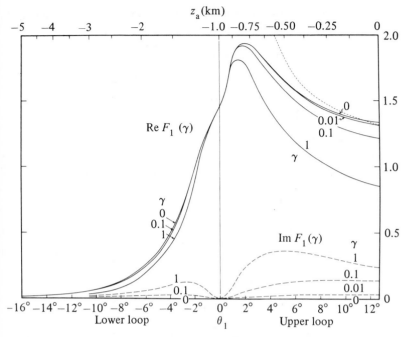

Fig. 10.2. Plots of $F_1(\gamma)$ from a numerical integration. Re $F_1(\gamma)$ is proportional to the loop contributions towards $\langle X^2 \rangle$. Re $F_1(0) = F_1^{\pm}$ corresponds to a single upward (θ_1 positive) or downward (θ_1 negative) loop. The short-dashed curve gives the apex approximation to F_1^+. (From Munk and Zachariasen, 1976.)

We shall also need the variance of $\dot{X} = dX/dt$; this is found by inserting ω^2 in (10.4.2) under the integral sign. The result is

$$\langle |\dot{X}^2| \rangle = \omega_i n_0 \langle \mu_0^2 \rangle \langle j^{-1} \rangle q^2 B R_0 F_2 \qquad (10.4.13)$$

where

$$R_0 F_2 = 8\pi^{-2} n_0^{-3} \int_0^R dx\, n^3 \log(n/\omega_i) f_2(\Delta), \qquad (10.4.14)$$

and

$$f_2(\Delta) = [\ln(n/\omega_i) - \tfrac{1}{2}\ln(\Delta^2/4) - \tfrac{1}{2}(\Delta^2+1)^{-\frac{1}{2}}$$
$$\cdot \ln\{[(\Delta^2+1)^{\frac{1}{2}}+1]/[(\Delta^2+1)^{\frac{1}{2}}-1]\}]/\ln(n/\omega_i). \qquad (10.4.15)$$

For the axial ray (where $f_2 = 1$) and near-surface rays we have, respectively,

$$\bar{F}_2 = 8\pi^{-2}(n_1/n_0)^3 R/R_0 \log n_1/\omega_i = 0.132 \ R/R_0$$

$$(10.4.16)$$

$$F_2^+ = 8\pi^{-2}(2\pi/3)^{\frac{1}{2}}(B|r|)^{\frac{1}{2}}(R_0)^{-1}(n_a/n_0)^3 \log (n_a/\omega_i).$$

$$(10.4.17)$$

F_2 is less peaked at the apex than F_1 and decreases monotonically with apex depth (Fig. 10.3).

Let us next turn to the quantity $\langle X^2 \rangle$. From (10.4.4) we find

$$\langle X^2 \rangle = -\langle \mu_0^2 \rangle \langle j^{-1} \rangle q^2 B R_0 G_1(R) \qquad (10.4.18)$$

where

$$G_1(R) = 4(\pi^2 \omega_i n_0^2 R_0)^{-1} \int_0^R dx \, n^3 f_1(\Delta) g(\beta, j_*).$$

$$(10.4.19)$$

Here we have defined

$$g(\beta, j_*) = \langle j^{-1} \rangle^{-1} \sum_{j=1}^{\infty} j^{-1} H(j) \exp(i\beta j^2) \qquad (10.4.20)$$

and β is defined by (10.4.5). Note that as $\beta \to 0$, $g(\beta, j_*) \to 1$ and thus $G_1 \to F_1$ and hence $\langle X^2 \rangle \to -\langle |X|^2 \rangle$.

An approximate analytic expression for $g(\beta, j_*)$ is

$$g(\beta, j_*) = \{\exp(-i\beta j_*^2) Ei[i\beta(j_*^2 + \tfrac{1}{4})] - Ei(i\beta/4)\} \ln^{-1}(4j_*^2 + 1),$$

and hence for small β,

$$G_1(R) = F_1(R) - 2j_*^2 [\pi \langle j^{-1} \rangle$$
$$\times (\pi j_* - 1)\omega_i n_0^2 R_0]^{-1} \int_0^R dx \, n^3 f_1(\Delta) |\beta| \dots \qquad (10.4.21)$$

For very long ranges where many loops are involved, we may simplify the integral in (10.4.19) as follows. First replace β by the many-loop long-range value β_{ML}:

$$\beta_{ML} = (\pi n/n_0 B)^2 [x(R-x)/qR](\tan^2 \theta/\delta) \equiv \alpha_{ML} \gamma(x)$$

$$(10.4.22)$$

where

$$\gamma(x) \equiv [x(R-x)/B^2 Rq]. \qquad (10.4.23)$$

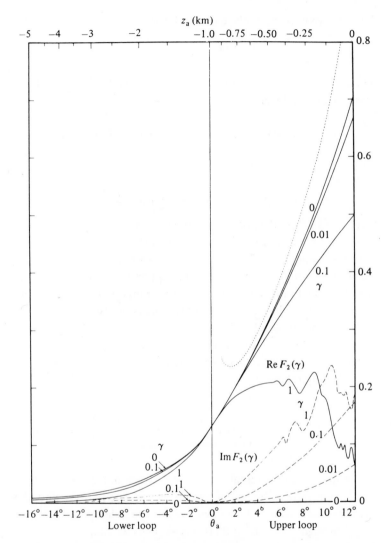

Fig. 10.3. Plots of $F_2(\gamma)$, proportional to the loop contributions towards $\langle X \rangle$. (From Munk and Zachariasen, 1976.)

Next note that γ varies rather little over one loop. Then the integral from 0 to R may be broken up into a sum of integrals over each of the loops. In any given loop, say the kth one, γ has very nearly the value $\gamma(x_k) \equiv \gamma_k$ where x_k is the position of the midpoint of the loop.

Thus we may write, in place of (10.4.19)

$$G_1 = \sum_{k=1}^{K^+} F_1^+(\gamma_k) + \sum_{k=1}^{K^-} F_1^-(\gamma_k)$$

where K^+ and K^- are the number of upper and lower loops respectively, and where

$$F_1^\pm(\gamma_k) = 4(\pi^2 \omega_i n_0^2 R_0)^{-1} \int_{x_k - \frac{1}{2}R^\pm}^{x_k + \frac{1}{2}R^\pm} dx \, n^3 f_1(\Delta) g(\beta_k, j_*)$$

(10.4.24)

with

$$\beta_k = \pi^2 (n/n_0)^2 (\tan^2 \theta / \delta) \gamma_k.$$

Here $x_k \pm \frac{1}{2} R^\pm$ are the positions of the two ends of the loop. We note that as $\gamma_k \to 0$, $F_1^\pm(\gamma_k) \to F_1^\pm$ as defined earlier.

The variation of $F_1^\pm(\gamma)$ with γ is shown in Fig. 10.2. Large deviations from $F_1^\pm$ begin to become apparent when γ is of order one. The maximum γ that occurs over a range R is $R/4B^2 q$; thus G_1 is not very different from F_1 until ranges of order $4B^2 q$.

Finally $\langle \dot{X}^2 \rangle$ is calculated in a way analogous to $\langle |\dot{X}|^2 \rangle$ defining G_2 through (10.4.19) with f_2 replaced by $f_2 g(\beta, j^*)$.

All the foregoing results and definitions may be summarized in the following relation:

$$\begin{bmatrix} -\langle X^2 \rangle \\ +\langle |X|^2 \rangle \\ -\langle \dot{X}^2 \rangle \\ +\langle |\dot{X}|^2 \rangle \end{bmatrix} = \langle \mu_0^2 \rangle \langle j^{-1} \rangle q^2 B R_0 \begin{bmatrix} G_1 \\ F_1 \\ \omega_i n_0 G_2 \\ \omega_i n_0 F_2 \end{bmatrix}.$$

(10.4.25)

The quantities of actual interest to us are not quite $\langle X^2 \rangle$ and $\langle |X|^2 \rangle$, but rather the mean-square phase fluctuations and the intensity fluctuations. These, we recall, are related to $\langle X^2 \rangle$ and $\langle |X|^2 \rangle$ through (10.1.18) and (10.1.19). Thus we find

$$\begin{pmatrix} \langle \phi^2 \rangle \\ \langle \iota^2 \rangle - \langle \iota \rangle^2 \end{pmatrix} = \langle \mu_0^2 \rangle \langle j^{-1} \rangle q^2 B R_0 \begin{pmatrix} \frac{1}{2} \\ 2 \end{pmatrix} (F_1 \pm \operatorname{Re} G_1).$$

(10.4.26)

For small β, $\operatorname{Re} G_1 \to F_1$ so that

$$\langle \phi^2 \rangle \to \langle \mu_0^2 \rangle \langle j^{-1} \rangle q^2 B R_0 F_1;$$

(10.4.27)

this is simply the conventional geometrical-optics expression for phase fluctuations. Note that in this limit $\langle |X|^2 \rangle = -\mathrm{Re}\,\langle X^2 \rangle = \langle \phi^2 \rangle = \Phi^2$. Intensity fluctuations, however, depend on the *difference* between F_1 and $\mathrm{Re}\,G_1$,

$$F_1 - \mathrm{Re}\,G_1 = 4(\pi^2 \omega_i n_0^2 R_0)^{-1} \int_0^R \mathrm{d}x\, n^3 f_1(\Delta)[1 - g(\beta, j_*)]$$

(10.4.28)

and thus vanish in the small β limit. Indeed, for small β, using (10.4.21) we find

$$\langle \iota^2 \rangle - \langle \iota \rangle^2 = 4\pi q \langle \mu_0^2 \rangle j_*^2 [(\pi j_* - 1)B\omega_i n_0^4]^{-1} \int_0^R \mathrm{d}x\, n^5 f_1(\Delta)|A^{-1}|.$$

(10.4.29)

We may make (10.4.29) more transparent by using our expression for L_V from (7.4.2):

$$\langle \iota^2 \rangle - \langle \iota \rangle^2 = \alpha 4 \langle \mu_0^2 \rangle \langle j^{-1} \rangle q^2 B(\pi^2 \omega_i n_0^2)^{-1} \int_0^R \mathrm{d}x\, n^3 f_1(\Delta)|qL_V^2 A|^{-1},$$

(10.4.30)

where

$$\alpha = \pi^3 j_*^2 \langle j^{-1} \rangle^{-1} (\pi j_* - 1)^{-3} \approx 1. \qquad (10.4.31)$$

Inspection of (10.4.6) to (10.4.9), (10.1.23), and (7.2.5) shows then that in the small β (geometrical-optics) limit; that is, when $\Lambda \ll 1$,

$$\langle \iota^2 \rangle - \langle \iota \rangle^2 = \alpha \Lambda \Phi^2 \approx \Lambda \Phi^2. \qquad (10.4.32)$$

On the other hand for very large β (large diffraction) $g(\beta, j_*)$ vanishes due to the fluctuating phase within the summation of (10.4.20), and we then find $F_1 - \mathrm{Re}\,G_1 \approx F_1$; hence

$$\langle \iota^2 \rangle - \langle \iota \rangle^2 = \langle \phi^2 \rangle \approx \tfrac{1}{2}\Phi^2 \qquad (10.4.33)$$

in the large diffraction ($\Lambda \gg 1$) limit.

It is also of interest to compute the spectra of phase and intensity fluctuations. For this purpose, we return to (10.4.2) and (10.4.4), but we do not now carry out the integral over $\mathrm{d}\omega$. For a given value of ω, we integrate over a ray path, keeping in mind that there is a complicated set of forbidden ray sections, depending on the value of ω relative to ω_L, n_a^- and n_a^+ (Fig. 10.4). For very low frequencies the ray is too steep to permit *stationary-phase interaction* with internal waves. For the high frequencies ω may exceed $n(z)$ along some

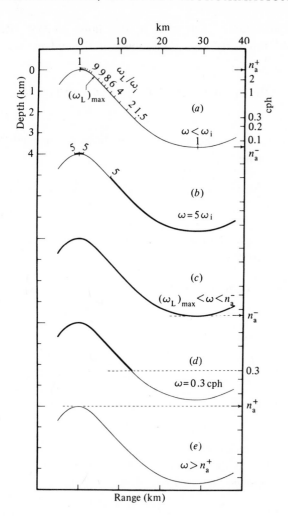

Fig. 10.4. Internal-wave contributions toward frequency spectra of acoustic phase and intensity come from 'permitted' sections of the ray path (heavy lines). There are no internal-wave contributions to frequencies less than the inertial frequency (a) and larger than the apex buoyancy frequency (e) because no such internal waves exist. The entire ray contributes toward the central band (c) between $(\omega_L)_{max}$ (typically $10\omega_i$) and the buoyancy frequency at the lower turning point (for deep rays (c) does not exist). For lower frequencies, the upper sections of the ray are too steep to permit 'stationary-phase interaction' (b), and high frequencies exceed buoyancy frequency of the deep ray section (d). (From Munk and Zachariasen, 1976.)

portions of the ray, and internal-wave solutions do not exist. The
phase and intensity spectra are given by

$$\begin{bmatrix} E_\phi(\omega) \\ E_\iota(\omega) \end{bmatrix} = (8/\pi^2)\langle\mu_0^2\rangle\langle j^{-1}\rangle q^2 B\omega_i n_0^{-2}\omega^{-3}(\omega^2 - \omega_i^2)^{\frac{1}{2}}$$

$$\cdot \int_0^R dx\, n^3(\omega^2 - \omega_L^2)^{-\frac{1}{2}} H(n - \omega)H(\omega - \omega_L)$$

$$\cdot \begin{bmatrix} \frac{1}{2}[1 + \mathrm{Re}\, g(\beta, j_*)] \\ 2[1 - \mathrm{Re}\, g(\beta, j_*)] \end{bmatrix}. \tag{10.4.34}$$

For the important range $\omega_L \ll \omega < n_a^-$ the entire integration path is
permitted and the spectra vary as ω^{-3}.

10.5. Comparison with numerical experiments

As a first application, and test, of the results we have obtained we
shall make a comparison with the set of numerical experiments of
Flatté and Tappert (1976). These consist of numerical solutions of
the parabolic wave equation in the same sound channel we have
discussed here, and with a sequence of internal-wave realizations
from a two-dimensional projection of the spectrum described in
Chapter 3.[†] The numerical experiments use an acoustic frequency
of 100 Hz, and propagate sound up to a range of 100 km; the
remaining parameters are the standard ones listed in Chapters 1 and
3. In all cases the acoustic transmitter is located on the sound axis, at
a depth of 1000 m. The receiver consists of a vertical array of
hydrophones, 700 m long, centered on the ray in question, which
allows an angular resolution of $1\frac{1}{2}°$.

Phase fluctuations

Solid lines in Fig. 10.5 show the results of the numerical experiment
for 128 realizations (to which one may assign a statistical error of
perhaps ±20°). The dotted lines are the predictions of the theory
outlined in §10.4, and specifically of (10.4.27). Evidently the
agreement is satisfactory. Overall magnitudes differ between theory
and experiment by about 20 to 30 per cent (except for the −1° ray)

[†] The numerical experiment uses the exact wavefunctions for the exponential model
ocean, whereas our analytical model is based on the corresponding WKB approximation.

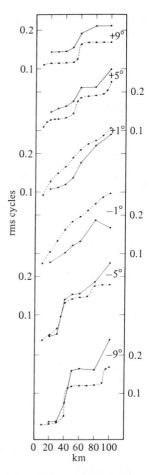

Fig. 10.5. Comparison of calculated (dotted lines) and computer-experiment (solid lines) rms phase fluctuations at 100 Hz as function of range for six rays with inclinations on the axis ranging between ±9°. Lines connect calculated values with no attempt at interpolation. (From Munk and Zachariasen, 1976.)

and the general shapes coincide as well. For the steep rays (±9° and to a lesser extent ±5°) the rms phase is nearly a step function of range, reflecting the fact that the major contribution to the integral in (10.4.7) comes when the rays cross an apex, and that there is little contribution while the ray is deep. The near axis rays (±1°), on the other hand, vary much more smoothly with range (nearly like $R^{\frac{1}{2}}$).

Table 10.1. *Rms intensity fluctuations in dB. The upper number in each entry is the theoretical value and the lower number the 'experimental' value.*

Ray angle	Range (km)						
	20	30	40	50	60	80	100
9°				1.00		1.21	1.34
				(0.73)		(0.82)	(1.02)
5°	1.10		1.12	1.39	1.31		
	(0.44)		(0.50)	(0.62)	(0.64)		
1°	0.79		1.03		1.31	1.57	1.80
	(0.34)		(0.41)		(0.60)	(0.74)	(0.95)
−1°	0.66		1.07		1.39	1.70	2.20
	(0.30)		(0.43)		(0.61)	(0.56)	(0.46)
−5°		0.38		0.94		1.19	1.29
		(0.41)		(0.61)		(1.02)	(0.84)
−9°	0.09	0.18	0.29			0.78	
	(0.06)	(0.07)	(0.31)			(0.63)	

Intensity fluctuations

Table 10.1 shows the rms intensity fluctuations for the same six rays, at various ranges. Since the numerical experiment makes use of a vertical beam former rather than a single hydrophone to select different rays, the theoretical calculations described in § 10.4 must be somewhat modified. In § 10.4 intensity fluctuations were calculated for a fixed receiver position; here we must calculate fluctuations for a fixed receiver angle but having a variable vertical position. This amounts to replacing the quantity A^{-1} in (10.4.29) by a different geometrical factor B^{-1}, defined to be the second derivative with respect to z of the optical path length from the transmitter to a receiver located at a fixed range and seeing a fixed vertical angle, rather than one located at a fixed range and height (See Chapter 7). This quantity has been evaluated numerically, and then (10.4.29) has been used, in order to obtain the theoretical values shown in the table.

For near-axial rays, B^{-1} becomes infinite when the receiver is located at the turning point of the rays. It is here that all rays are parallel; this is the analogue of a caustic for a beam-former receiver. As we have already remarked in § 10.3, when the receiver is placed near a caustic, our approximate expressions (10.4.4) fail. We are therefore able to compare our calculated values with the computer ones only if we avoid placing the receiver near a (beam-former type of) caustic. For near-axis rays, where we can use (7.1.19), these occur at ranges $R \approx (n + \frac{1}{2})\pi/K$; since $\pi/K \approx R_0 \approx 20$ km for near-axis rays, there are caustics at ranges of 10, 30, 50, ... km. For off-axis rays, the positions of caustics must be determined numerically. The missing entries in Table 10.1 are due to these caustics.

The agreement between theory and computer experiment is satisfactory for the off-axis rays but for near-axis rays the theory seems to overestimate the size of the fluctuations; in particular, the $\pm 1°$ rays are predicted to have fluctuations that are larger by a factor of two to three than the computer experiment shows. This discrepancy may be due to any one, or a combination of, four factors: (1) the use of the linear approximation to the dispersion relation; (2) that the WKB approximation underlying the theory does not include the reduction in vertical displacement near the boundaries; (3) the two-dimensional character of the computer experiment; (4) the failure of the expansion in powers of acoustic wavelength to the ratio of vertical correlation length.

CHAPTER 11

PROPAGATION THROUGH A
SINGLE UPPER TURNING POINT

In this chapter we will study propagation through a single upper turning point. This problem turns out to be equivalent to the so-called phase-screen problem which was elegantly solved by Mercier (1962), and further elucidated by Salpeter (1967). We will describe both strong and weak scattering with the formalism introduced here, but the purpose of the chapter is to begin the solution for sound propagation in the presence of strong scattering, in particular where the sound-speed fluctuations break deterministic rays into several sporadic rays (the saturated regimes of Fig. 8.4). In Chapter 12, the mathematics contained in the phase-screen problem will be combined with the path-integral formalism introduced in Chapter 5 to solve the general problem of the saturated regime.

We begin by setting up the problem using the formulas of geometrical optics (§ 11.1), following with a qualitative discussion of the various regions of the range–frequency diagram (§ 11.2). We then concentrate on the saturated regime by emphasizing the cases where there are many stationary-phase points in the expression for the pressure, and we give expressions for phase and intensity fluctuations, as well as the probabilities of sporadic multipath (§ 11.3 and § 11.4). Sections 11.3 and 11.4 calculate the same quantities (statistics of pressure fluctuations); § 11.3 gives a direct, more intuitive method of calculating, while § 11.4 uses a method that can more easily be generalized (the generalization is carried out in Chapter 12). Correlations between waves of different frequency are studied in § 11.5.

11.1. Setting up the problem

We assume a source at $(0, 0, z_s)$ and a receiver at (R, y, z_r) where $y/R \ll 1$, and suppose that there is an unperturbed ray $z_{ray}(x)$ with a

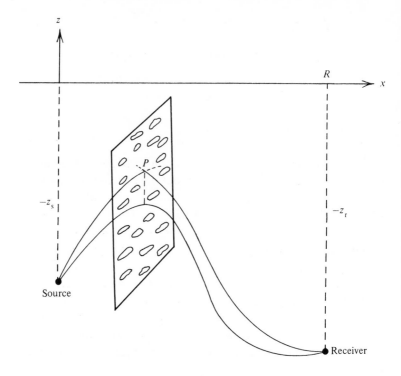

Fig. 11.1. Schematic of the phase-screen approximation. All scattering is assumed to take place in the plane of the upper turning point.

single upper turning point which connects $(0, 0, z_s)$ to $(R, 0, z_r)$. The upper turning point at $(x_a, 0, z_a)$ is assumed to be well above z_s and z_r (Fig. 11.1).

The vertical coordinate of the receiver z_r turns out to be a rather inconvenient variable. It is more convenient to use the height of the turning point $\hat{z}(z_r)$. For z_r close to some fixed point z_0 we can effect this change of variables by setting $v = \hat{z} - \hat{z}(z_0) = d\hat{z}/dz_r(z_r - z_0)$ and writing our equations in terms of v. One can express $d\hat{z}/dz_r$ in terms of the phase curvature $A \equiv A(\hat{x}_0)$ where $A(x)$ is defined in Chapter 7, and $\hat{x}_0$ is the horizontal position of the turning point for the ray going from $(0, 0, z_s)$ to $(R, 0, z_0)$.

The angle of a ray is $\theta(x) = dz_{ray}/dx$ and r is the radius of curvature of the ray at the turning point. The relation between v and

z_r is then

$$v = (z_r - z_0)/rA\theta(R). \tag{11.1.1}$$

Given this ray geometry and the sound-speed correlation function, described in Chapter 6, we wish to find an expression for the wavefunction ψ under the following conditions:

(i) Most of the scattering will take place at the upper turning point.

(ii) Upon passage through the upper turning point the phase of the wave is modulated but there is little change in intensity.

(iii) The phase modulation can be computed using geometrical optics.

These assumptions are valid if horizontal propagation over a distance $(rL_V)^{\frac{1}{2}}$ around the turning point remains in the geometric-optics regime. This is true in essentially all cases of practical interest.

Our assumptions allow us to apply Huygens' principle to our configuration in the following way. An unperturbed ray is followed to some position (P) in the $y-z$ plane containing the upper turning point. At this point the sound-speed fluctuation is assumed to add a phase (and thereby bend the ray), whereupon an unperturbed ray is followed to the receiver (Fig. 11.1). The wavefunction at the receiver is the sum of all such contributions where P may take any position in the plane of the upper turning point (this plane is called the 'phase screen'). The resulting expression for ψ is:

$$\psi(y, v, t) = \left(\frac{qA}{2\pi i}\right)^{\frac{1}{2}} \int_{-\infty}^{\infty} \exp\left[\tfrac{1}{2}iqA(v'-v)^2 + i\phi(v', y, t)\right] dv' \tag{11.1.2}$$

where the term containing the phase curvature, A, represents the effect of the sound channel geometry; and ϕ represents the sound-speed fluctuations. (One sees that if $\phi = 0$ then one has $\psi = \psi_0$, that is, there is no perturbation.)

Even though the phase addition ϕ is assumed to act only at the upper turning point, we take for its magnitude the integral of the sound-speed fluctuation $\delta C/C$ along the entire path (most of the contribution to the integral will come from near the upper turning point). Thus ϕ is a Gaussian random variable with zero mean and a variance given by

$$\langle \phi(v, y, t)\phi(v', y', t')\rangle$$

$$= q^2 \int_{-\infty}^{\infty} dx \int_{-\infty}^{\infty} dx' \, \rho\left(x - x', \frac{\hat{x}_0}{R}(y - y'), v - v' - \frac{x^2 - x'^2}{2r}, t - t'; \hat{z}\right.$$

$$\left. - \frac{x^2 + x'^2}{4r}\right). \qquad (11.1.3)$$

A glance at (7.2.1) reveals that ϕ is in fact a generalization of our fluctuation parameter Φ such that

$$\langle \phi^2(0, 0, 0)\rangle = \Phi^2. \qquad (11.1.4)$$

Now the quantity v is a vertical displacement. The size of this vertical displacement relative to the vertical correlation length L_V governs the fluctuation effects, so a better (dimensionless) variable will be $\xi = v/L_V$. Remembering from Chapter 7 that our other fluctuation variable Λ is defined so that $\Lambda = |qAL_V^2|^{-1}$, we have $-qA(v' - v)^2 = (\xi' - \xi)^2/\Lambda$. Note that A is negative in this single-apex case. Thus Λ governs the size of the first term in the exponential of (11.1.2) while ϕ governs the size of the second term. To bring out the ϕ effect more clearly we define a new random variable u such that

$$u(\xi, y, t) = \frac{\phi(v, y, t)}{\Phi}. \qquad (11.1.5)$$

Thus u has the same shape as ϕ, but is normalized such that $\langle u^2(0, 0, 0)\rangle = 1$. We can now express the reduced wavefunction in a canonical form;

$$\psi(\xi, y, t) = (i/2\pi\Lambda)^{\frac{1}{2}} \int_{-\infty}^{\infty} \exp\left[-\frac{1}{2}i(\xi' - \xi)^2/\Lambda + i\Phi u(\xi', y, t)\right] d\xi'.$$

$$(11.1.6)$$

This equation applies to a particular source–receiver geometry with a single upper turning point. If there were no sound-speed fluctuations the pressure would be given by $\psi_0(\xi)$ where ξ represents a small vertical displacement around the receiver. With sound-speed fluctuations, two normalized variables enter; ξ a vertical displacement relative to L_V, and u, a random variable containing the shape of the sound-speed correlation function. The sizes of the two terms are governed by the fluctuation parameters:

Λ, relating the size of a Fresnel zone relative to L_V; and Φ, the rms phase fluctuation calculated in geometrical optics.

It will be useful to define a correlation function, μ, for u such that

$$\mu(\xi - \xi', y - y', t - t') \equiv \langle u(\xi, y, t)u(\xi', y', t')\rangle \qquad (11.1.7)$$

evidently, from (11.1.4),

$$\mu(0, 0, 0) = 1. \qquad (11.1.8)$$

The Fourier transform $\tilde{\mu}$ is given by

$$\tilde{\mu}(\lambda) = \frac{1}{2\pi} \int_{-\infty}^{\infty} e^{-i\lambda\xi} \mu(\xi, 0, 0)\, d\xi \qquad (11.1.9)$$

and for large λ it behaves like

$$\tilde{\mu}(\lambda) = \frac{C_\mu}{|\lambda|^{p+1}} \qquad (11.1.10)$$

where C_μ is a constant of order unity and $p \approx 2$. (This result can be derived from the behavior of ρ given in Chapter 6.)

Equations (11.1.9) and (11.1.10) can be combined to show that for small ξ

$$1 - \mu(\xi, 0, 0) = \frac{1}{2} \frac{|\xi|^p - \xi^2}{2 - p}. \qquad (11.1.11)$$

For internal waves $p \approx 2$ and $1 - \mu(\xi, 0, 0) \sim \frac{1}{2}\xi^2|\ln \xi|$. In this chapter we will retain p as an arbitrary number between 1 and 4, allowing us to exhibit the effect of different values of p on the statistics of ψ.

The variables y and t appear only as parameters in (11.1.6) and will generally be suppressed. The reason why there is no integral over a horizontal coordinate is the following. Because L_H/L_V is so large we can always treat y variations using geometric optics. That is, had we started with an integral over y' we could have evaluated the integral by the stationary-phase method. There would have been a unique stationary-phase point at $y' = yx_a/R$ and the result would be just (11.1.6).

11.2. Regions in Λ–Φ space

Let us explore the regions in Λ–Φ space for the single-apex geometry (11.1.6). See Fig. 8.3 for the definitions of the regions. In the first Born approximation region, Φ is small and we may expand the integrand to obtain $\psi \approx 1 + iX$ where

$$X(\xi) = \Phi(i/2\pi\Lambda)^{\frac{1}{2}} \int_{-\infty}^{\infty} \exp\left[-\tfrac{1}{2}i(\xi' - \xi)^2/\Lambda\right]u(\xi')\,d\xi'.$$

(11.2.1)

In the overlapping geometrical-optics regime, where Λ is small, the integral can be done by the stationary-phase method yielding $X(\xi) \approx i\Phi u(\xi)$. In the region where Λ is large the complete Born approximation is necessary.

In the saturated regime the phase of the integrand is necessarily large because Φu is large. We can therefore contemplate doing the integral by a stationary-phase approximation. Setting $\xi = 0$ for simplicity the stationary-phase points are the values of ξ' for which $\tfrac{1}{2}(\xi')^2/\Lambda + \Phi u(\xi')$ is at a maximum or a minimum. Concentrating on the second term for the moment we note that in general u is a very wiggly function with many local maxima and minima. Not all of these maxima and minima can produce meaningful stationary-phase points. For two stationary-phase points to be distinct, it is necessary that the phase of the integrand at the two points differ by at least $\pi/2$, i.e., order unity. This is not the case for maxima and minima produced by the small high-frequency wiggles in u. We will get around this problem by dividing u into a low-frequency part u_1 and a small high-frequency part u_2 according to

$$u(\xi') = u_1(\xi') + u_2(\xi')$$
$$= \int_{|\lambda| < \lambda_c} e^{i\lambda\xi'}\tilde{u}(\lambda)\,d\lambda + \int_{|\lambda| > \lambda_c} e^{i\lambda\xi'}\tilde{u}(\lambda)\,d\lambda. \qquad (11.2.2)$$

The dividing point will be chosen such that $\Phi\langle(u_2)^2\rangle < 1$. For our power-law spectrum this can be accomplished by taking $\lambda_c = (\Phi)^{2/p}$. Since u_2 is small we can write

$$\psi(0) = (i/2\pi\Lambda)^{\frac{1}{2}} \int_{-\infty}^{\infty} \exp\left[-\tfrac{1}{2}i(\xi')^2/\Lambda + i\Phi u_1(\xi')\right][1 + i\Phi u_2(\xi')]\,d\xi'$$

(11.2.3)

where the first term is large and is our main concern. The second term proportional to u_2 is a small correction.

The main term in (11.2.3) contains only u_1 and the integral can be done by the stationary-phase method. The points of stationary phase are determined by

$$-\xi' + \Phi\Lambda \, d/d\xi' \, u_1(\xi') = 0 \qquad (11.2.4)$$

and by construction all solutions to (11.2.4) represent distinct stationary-phase points.

There are oscillations in u_1 with wavelengths ranging from order unity (spatial scales $\sim L_V$) to order $\Phi^{-2/p}$ (spatial scales $\sim \Phi^{-2/p} L_V$) and we would like to examine the stationary-phase points generated by oscillations of various scales. We therefore define u_l for $1 > l > \Phi^{-2/p}$ by

$$u_l(\xi') = \int d\lambda \, e^{i\lambda\xi'} \tilde{u}(\lambda) \, d\lambda \qquad (11.2.5)$$

where the integral goes over values of λ such that $l^{-1} < |\lambda| < \Phi^{2/p}$. The magnitude of u_l is of order $l^{p/2}$ and $d/d\xi$ applied to u_l is of order l^{-1}. Now let $\xi_{max}(l)$ be the largest value of $|\xi'|$ for which (11.2.4) can be solved when u_1 is replaced by u_l: it is

$$\xi_{max}(l) \sim \Phi\Lambda l^{(\frac{1}{2}p - 1)}. \qquad (11.2.6)$$

Note that for $p > 2$, ξ_{max} decreases as l decreases which means that the stationary-phase points produced by the smaller scales lie very close to the unperturbed ray going through $\xi = 0$. Just the opposite happens for $p < 2$. In this case the smaller scales produce stationary-phase points which are far away from $\xi = 0$. The statistics of ψ in the partially saturated region are different for $p > 2$ and $p < 2$. The ultimate reason for this is the curious geometrical phenomenon which we see here.

If $\xi_{max}(l)$ is less than l for all scale sizes then it is clear that (11.2.4) will have only one solution which will be near $\xi' = 0$. One can readily verify that the condition that there be a single stationary-phase point is $\Phi^{4/p}\Lambda < 1$ which defines one boundary of the unsaturated region, the other being $\Phi > 1$. Thus in the unsaturated region there is a single stationary-phase point and (11.1.6) reduces to

$$\psi(\xi) = \exp[i\Phi u(\xi)] \qquad (11.2.7)$$

where we have taken $\xi \neq 0$ which will move the stationary-phase point to $\xi' = \xi$. The parameter Λ is necessarily small in the geometrical-optics region and a formal expansion of the integrand will give a Born approximation $-iX(\xi) = \Phi u(\xi)$. Thus the expression $\psi = e^X$ is valid throughout the unsaturated region. This is the Rytov approximation of Chapter 10.

In the saturated regions there will be multiple stationary-phase points. For $p < 4$ an estimate of the number of such points is $\Phi^{2/p}\xi_{max}(\Phi^{-2/p}) = \Phi^{4/p}\Lambda$. The integral is then

$$\psi(\xi) = \sum_i A_i \, e^{i\phi_i} + \text{(small background)} \qquad (11.2.8)$$

where the ϕ_i are the phases of the integrand at the stationary-phase points, the A_i are constants determined by the curvature of the phase at the stationary-phase point and the 'small background' comes from the second term in (11.2.3) plus corrections to the stationary-phase approximation in the first term. Equation (11.2.8) clearly shows multipath propagation. There is a finite, *discrete* set of ray paths plus a small 'scattered wave'. In the saturated regions a ray approximation is therefore valid. It is not a simple ray approximation however, since the fluctuations are producing multiple propagation paths. The meaning of all this can be easily understood if we think of an experiment with a pulsed source. The receiver will detect a finite, discrete series of sharp pulses, one for each term in the sum over i and in addition there will be a small background signal spread over a continuum of arrival times.

Let us now find the boundary between the partially and fully saturated regions. If the maximum over l of $\xi_{max}(l)$ is one or larger, then the different propagation paths will tend to be uncorrelated. One finds that the maximum over l of $\xi_{max}(l)$ is greater than unity if $\Phi^{2/p}\Lambda > 1$ for $p < 2$ or $\Phi\Lambda > 1$ for $p > 2$ which defines the boundary. In the fully saturated region the paths can be separated by a vertical correlation length L_V while in the partially saturated region they all lie within L_V of the unperturbed ray. It is also interesting to compute the maximum scale length l_{max} which can produce multiple paths. Setting $\xi_{max}(l_{max}) = l_{max}$ we find $l_{max} = \Phi^{[2/(4-p)]}\Lambda$. Thus another characteristic of these regions is that in the partially saturated region only the small scales $l \ll 1$ are contributing to multipathing while in full saturation all scales are contributing.

The fluctuating part of the travel time along a path is $t_i = \sigma^{-1}(-\xi_i^2/2\Lambda + \Phi u(\xi_i))$. In full saturation the first term $-\xi_i^2/2\sigma\Lambda$ dominates and the fluctuations in travel time are large compared to the values calculated using the Rytov approximation. Also the differences in travel times $t_i - t_j$ are comparable to the wander in t_i for a given path as was shown in Fig. 8.5. In partial saturation for any particular t_i the second term $\Phi u(\xi_i)$ dominates, but travel time differences are dominated by the first term. This makes the spacing between arrivals small compared to the wander in arrival time of the group as a whole.

11.3. Sound fluctuations in the presence of micromultipath

We will now work out the statistics of ψ using a simple intuitive method which emphasizes the fact that there are many stationary-phase points, i.e. propagation paths, in the saturated regions. We do the stationary-phase integral *before* finding expectation values, so that ψ is always a sum of separate ray contributions.

The phases are large and random so that

$$\langle\psi\rangle = \left\langle \sum_i A_i\, e^{i\phi_i}\right\rangle = 0 \tag{11.3.1}$$

and

$$\langle\psi^2\rangle = \left\langle \sum_{ij} A_iA_j\, e^{i(\phi_i+\phi_j)}\right\rangle = 0. \tag{11.3.2}$$

However, for the intensity $I = |\psi|^2$ we find

$$\langle I\rangle = \left\langle \sum_{ij} A_iA_j\, e^{i(\phi_i-\phi_j)}\right\rangle = \left\langle \sum_i A_i^2\right\rangle. \tag{11.3.3}$$

If p is greater than 2, the average of $\sum_i A_i^2$ can be computed by observing that

$$\sum_k A_k^2 = \sum_k |1 - \Phi\Lambda u''(\xi_k)|^{-1} = \int_{-\infty}^{\infty} \delta[\xi - \Phi\Lambda u'(\xi)]\, d\xi \tag{11.3.4}$$

where the ξ_k are the stationary-phase points. Then by the rules of Gaussian statistics

$$\left\langle \sum_k A_k^2\right\rangle = \int_{-\infty}^{\infty} \langle\delta[\xi - \Phi\Lambda u'(\xi)]\rangle\, d\xi = \int_{-\infty}^{\infty} P_1(\xi)\, d\xi = 1 \tag{11.3.5}$$

where

$$P_1(\xi) = [2\pi\langle\xi^2\rangle]^{-\frac12} \exp[-\tfrac12\xi^2/\langle\xi^2\rangle] \tag{11.3.6}$$

with

$$\langle\xi^2\rangle = (\Phi\Lambda)^2|\mu''(0)|. \tag{11.3.7}$$

The function $P_1(\xi)$ can be interpreted as the probability of finding a stationary-phase point at ξ (in other words, a ray). Note that $\langle \xi^2 \rangle$ is small for partial saturation, and large for full saturation, consistent with the argument above.

Let us now calculate some higher moments. Because of the random phases, $\langle \psi^{*n} \psi^m \rangle$ will vanish unless $n = m$. The simplest nonvanishing object is $\langle |\psi|^4 \rangle \equiv \langle I^2 \rangle$ which is given by

$$\langle I^2 \rangle = \sum_{ijkl} A_i A_j A_k A_l \exp\left[i(\phi_i - \phi_j + \phi_k - \phi_l)\right]$$

$$= 2\left\langle \sum_{kj} A_k^2 A_j^2 \right\rangle$$

$$= 2 \int_{-\infty}^{\infty} P_2(\xi_1, \xi_2) \, d\xi_1 \, d\xi_2 \tag{11.3.8}$$

where the factor of two appears because the phases will cancel for either $i = j$, $k = l$ or $i = l$, $k = j$ and the quantity P_2 defined by

$$P_2(\xi_1, \xi_2) = \langle \delta[\xi_1 - \Phi\Lambda u'(\xi_1)]\delta[\xi_2 - \Phi\Lambda u'(\xi_2)]\rangle \tag{11.3.9}$$

is a measure of the probability that ξ_1 and ξ_2 will both be stationary-phase points. In full saturation an examination of this function shows that in the integral in the last line of (11.3.9) we can approximate $P_2(\xi_1, \xi_2)$ by $P_1(\xi_1)P_1(\xi_2)$ and $\langle I^2 \rangle = 2$. In partial saturation with $p \approx 2$ one can also verify that $P_2(\xi_1, \xi_2) \approx P_1(\xi_1)P_1(\xi_2)$ with an error of order $[\ln(\Phi^2\Lambda)]^{-1}$. However, in partial saturation with $p > 2$ an explicit calculation shows that P_2 cannot be factorized and the locations of the stationary-phase points are correlated.

One can generalize the definition of P_2 to $P_n(\xi_1, \ldots, \xi_n)$ in the obvious way. In full saturation, P_n always factorizes according to $P_n(\xi_1, \ldots, \xi_n) = \prod_{i=1}^{n} P_1(\xi_i)$ and the locations of the stationary-phase points are independent. This is also true in partial saturation for $p < 2$ and with modest accuracy for $p \approx 2$.

In general

$$\langle I^n \rangle = n! \int_{-\infty}^{\infty} P_n(\xi_1 \ldots \xi_n) \, d\xi_1 \ldots d\xi_n \tag{11.3.10}$$

where the $n!$ comes from the number of ways that a phase can cancel in a sum like that in (11.3.8). When P_n factorizes we have

$$\langle I^n \rangle = n!\langle I \rangle^n. \tag{11.3.11}$$

We have now shown that in the fully saturated region, and in the partially saturated region with $p \lesssim 2$, that

$$\langle \psi^{*n} \psi^m \rangle = n! \langle I \rangle^n \delta_{nm}. \tag{11.3.12}$$

This means that the real and imaginary parts of ψ are independent Gaussian random variables with zero mean and a variance given by $\langle I \rangle = 1$.

Now suppose that the fluctuations depend on time and that $u(t', \xi) = u(t, \xi) + \delta u(\xi)$ where $\delta u \sim (t' - t) \partial_t u$ is small. The change in phase for the kth propagation path is then $\phi_k(t') = \phi_k(t) + \delta \phi_k$ where $\delta \phi_k = \Phi \delta u(\xi_k)$ and can be of order unity because Φ is large. For a small $\delta u \sim \Phi^{-1}$, changes in the amplitudes A_k are not important. We will take δu to be independent of u so that there is no correlation between δu and the A_ks and ξ_ks. In the internal-wave model $\langle u(\xi, t) \partial_t u(\xi', t) \rangle$ vanishes and this assumption is justified. It is also valid for a time dependence due to displacement of the inhomogeneities by a horizontal current.

With these assumptions we find for the second-order correlation

$$\langle \psi^*(t') \psi(t) \rangle = \left\langle \sum_{kj} A_k A_j \exp \left[i(\phi_j - \phi_k - \delta \phi_k) \right] \right\rangle$$

$$= \left\langle \sum_k A_k^2 \exp \left[-i \delta \phi_k \right] \right\rangle$$

$$= \exp \left[-\tfrac{1}{2} \Phi^2 \delta^2 \mu(0) \right] \tag{11.3.13}$$

where

$$\delta^2 \mu(\xi - \xi') \equiv \langle \delta u(\xi) \, \delta u(\xi') \rangle = \langle (\partial_t u)^2 \rangle (t - t')^2. \tag{11.3.14}$$

Noting that $\Phi^2 \delta^2 \mu(0) = D(t' - t)$, (11.3.13) becomes identical to (8.4.5), thus proving that relationship. To see if ψ is Gaussian, compute

$$\langle I(t') I(t) \rangle = \sum_{jklm} \langle A_k A_j A_l A_m \exp \left[i(\phi_j - \phi_m + \phi_l + \delta \phi_l - \phi_k - \delta \phi_k) \right] \rangle$$

$$= \sum_{jl} \langle A_j^2 A_l^2 \rangle + \sum_{jl} \langle A_j^2 A_l^2 \exp \left[i(\delta \phi_j - \delta \phi_l) \right] \rangle$$

$$= \tfrac{1}{2} \langle I^2 \rangle + \int_{-\infty}^{\infty} \int P_2(\xi, \xi') \exp \left\{ -\Phi^2 [\delta^2 \mu(0) \right.$$
$$\left. - \delta^2 \mu(\xi - \xi')] \right\} \, d\xi \, d\xi'. \tag{11.3.15}$$

In full saturation most of the contribution to the integrals over ξ and

ξ' comes from regions where $|\xi - \xi'| > 1$ and $\delta^2 \mu (\xi - \xi')$ is small yielding

$$\langle I(t)I(t') \rangle = \langle I \rangle^2 \{1 + \exp[-\Phi^2 \delta^2 \mu (0)]\}$$
$$= \langle I \rangle^2 \{1 + \exp[-D(t - t')]\}. \qquad (11.3.16)$$

This relation proves the Gaussian character of ψ in the fully saturated region. In the partially saturated region, the integral is dominated by small ξ and the behavior of the intensity correlation depends critically on the behavior of $\delta^2 \mu (\xi)$ for small ξ. For the internal-wave-induced fluctuations $\delta^2 \mu (\xi)$ has the same behavior at small ξ as does $\mu (\xi)$. In this case one cannot neglect $\delta^2 \mu (\xi - \xi')$ and correlations in times are not Gaussian. The evaluation of $\langle I(t)I(t') \rangle$ for internal waves in the partially saturated region will be given later. For a time dependence which is due to displacement of the inhomogeneities by a horizontal current, the small ξ behavior of $\delta^2 \mu$ is much different. In this case the spectrum of $\delta^2 \mu$ behaves like $|\lambda|^{1-p}$ rather than $|\lambda|^{-1-p}$ and as a function of $\xi - \xi'$, $\delta^2 \mu$ vanishes much more rapidly than μ. When $p < 2$ or $p \approx 2$ and when the Taylor hypothesis is valid $\delta^2 \mu (\xi)$ can be neglected in the partially saturated region and the statistics in time become Gaussian. We will not discuss the case $p < 2$ in partial saturation which is more complicated.

Correlations in the transverse horizontal direction y are similar to correlations in time when the Taylor hypothesis is valid. They are Gaussian in the fully saturated region, and in the partially saturated region for $p < 2$ or $p \approx 2$.

11.4. A better method of calculating sound fluctuations in the presence of micromultipath

We are now going to work through some of the statistics of ψ using a better calculation scheme which can be carried over to the path integral calculations in later sections. In this section we do *not* carry out the stationary-phase integral before averaging.

From the integral representation in (11.1.6) one can compute directly that

$$\langle \psi \rangle = \exp(-\tfrac{1}{2}\Phi^2) \approx 0 \qquad (11.4.1)$$

for large Φ. For the average of ψ^2 we find

$$\langle\psi^2\rangle = (i/2\pi\Lambda)\int_{-\infty}^{\infty}\exp\{-\tfrac12 i(\xi^2+\xi'^2)/\Lambda - \Phi^2[1+\mu(\xi-\xi')]\}\,d\xi\,d\xi'.$$

$$(11.4.2)$$

The integrand is of order $\exp(-\Phi^2)$ in all regions of $\xi-\xi'$ space so that $\langle\psi^2\rangle$ like $\langle\psi\rangle$ is vanishingly small. In general, any correlation of the form $\langle(\psi^*)^m(\psi)^n\rangle$ with $m\neq n$ will be exponentially small.

The simplest nonvanishing correlation is

$$\langle\psi^*(\xi', y', t')\psi(\xi, y, t)\rangle$$

$$= (2\pi\Lambda)^{-1}\int_{-\infty}^{\infty}\exp\{-\tfrac12 i(\xi_1^2-\xi_2^2)/\Lambda - \Phi^2$$

$$\cdot[1-\mu(\xi_1-\xi_2+\xi-\xi', y-y', t-t')]\}\,d\xi_1\,d$$

$$= \exp\{-\Phi^2[1-\mu(\xi-\xi', y-y', t-t')]\}$$

$$= \exp[-\tfrac12 D(\xi-\xi', y-y', t-t')]$$

$$(11.4.3)$$

where the integral is effected by changing variables to $\xi_1+\xi_2$ and $\xi_1-\xi_2$ and integrating over $\xi_1+\xi_2$, which produces a δ-function of $\xi_1-\xi_2$. Note that (8.4.5) has again been derived.

Next consider the correlation

$$\langle I\rangle^2 = \frac{1}{(2\pi\Lambda)^2}\int_{-\infty}^{\infty}d^4\xi\,\exp\left[\tfrac12 i\sum_{j=1}^{4}(-1)^j\xi_j^2/\Lambda - M\right] \quad (11.4.4)$$

where

$$M = \tfrac12\Phi^2\sum_{i,j=1}^{4}(-1)^{i+j}\mu(\xi_i-\xi_j). \quad (11.4.5)$$

For large Φ the factor $\exp(-M)$ is very small except in two regions, given by

(a) $|\xi_1-\xi_2|\lesssim\Phi^{-2/p}$, $|\xi_3-\xi_4|\lesssim\Phi^{-2/p}$ with $|\xi_1-\xi_3|$, $|\xi_2-\xi_4|$ arbitrary

and

(b) $|\xi_1-\xi_4|\lesssim\Phi^{-2/p}$, $|\xi_2-\xi_3|\lesssim\Phi^{-2/p}$ with $|\xi_1-\xi_2|$ and $|\xi_3-\xi_4|$ arbitrary

where it has been assumed that $p\leq 2$: for $p>2$, $\Phi^{-2/p}$ should be replaced by Φ^{-1}. In region (a) the oscillating phase $\exp[\tfrac12 i\sum_{j=1}^{4}(-1)^j\xi_j^2/\Lambda]$ cuts off the integral for $|\xi_1-\xi_3|$ or $|\xi_2-\xi_4|$ of order $\Lambda\Phi^{2/p}$ if $p < 2$ or $\Lambda\Phi$ if $p > 2$. There is a similar cutoff in region (b) and the total integration volume in either region is $\Lambda\Phi^{2/p} \times \Phi^{-2/p} = \Lambda$. This is large compared to the volume $\Phi^{-4/p}$ (or Φ^{-2}) of the region where (a) and (b) overlap. We may therefore replace the integral in (11.4.4) by the sum of two integrals taken separately over regions (a) and (b).

In most of region (a) $|\xi_1 - \xi_3|$ and $|\xi_2 - \xi_4|$ will be of order $\Lambda\Phi^{2/p}$ which in full saturation is large compared to unity. There can then be no correlation between the pairs of points (ξ_1, ξ_2) and (ξ_3, ξ_4) and it is valid to approximate M by

$$M \approx \tfrac{1}{2}\Phi^2[1 - \mu(\xi_1 - \xi_2) + 1 - \mu(\xi_3 - \xi_4)] \qquad (11.4.6)$$

in region (a) and

$$M \approx \tfrac{1}{2}\Phi^2[1 - \mu(\xi_1 - \xi_4) + 1 - \mu(\xi_2 - \xi_3)] \qquad (11.4.7)$$

in region (b). The integral in (11.4.4) then reduces to the square of the integral for $\langle I \rangle$ and we have our previous result that $\langle I^2 \rangle = 2\langle I \rangle^2$. For partial saturation $\Lambda\Phi^{2/p}$ is less than unity and the above argument does not apply. Nevertheless, for $p < 2$ the fractional power behavior of μ at small ξ has the effect of making (11.4.6) and (11.4.7) valid in partial saturation. For $p > 2$ (11.4.6) and (11.4.7) are not valid in the partially saturated case and as noted before $\langle I^2 \rangle \neq 2\langle I \rangle^2$.

The generalization to $\langle (\psi^*\psi)^n \rangle = \langle I^n \rangle$ is straightforward. There will now be $n!$ important regions in the integration volume corresponding to the $n!$ ways of pairing ψs with ψ^*s. For full saturation or partial saturation with $p < 2$ the integral in each of the $n!$ regions can be approximated by the nth power of the integral for $\langle I \rangle$ with the result that $\langle I^n \rangle = n!\langle I \rangle^n$ as expected.

It is interesting to see how the present method is related to the 'micropath method' used before. In a given one of the $n!$ regions we can take the pairing to be $(\xi_1, \xi_{n+1}), (\xi_2, \xi_{n+2}), \ldots, (\xi_n, \xi_{2n})$. Now make a change of variables to $\eta_k = \xi_k - \xi_{n+k}$ and $\bar{\xi}_k = \tfrac{1}{2}(\xi_k + \xi_{n+k})$. Assuming $p > 2$ for notational simplicity, the ηs are restricted to small values $\leq \Phi^{-1}$ and the analog of M can be expanded around $\eta_k = 0$. It is sufficient to keep the first nonvanishing terms: they are quadratic in the ηs. The Gaussian integral over the ηs yields a function of the $\bar{\xi}$s which is precisely $P_n(\bar{\xi}_1, \ldots, \bar{\xi}_n)$. The remaining integral over the $\bar{\xi}$s then gives $\langle I^n \rangle$ as in (11.3.10). When the analog of (11.4.6) and (11.4.7) holds it is clear that P_n factorizes. Previously we were doing stationary-phase integrals and then averaging. Now when we average before integrating the approximation has become an integration by steepest descents. In fact the correspondence

stationary phase integration before averaging $\leftrightarrow$

steepest descent integration after averaging

is general and will continue to hold in the next chapter where similar methods are applied to the path integral.

Higher-order correlations in ξ, y, t can be handled in a similar way. There are always the $n!$ important regions in the integration volume. In each region the integral usually factorizes into the product of n integrals for second-order correlations yielding Gaussian statistics for ψ. The exceptional case being correlations in time in the partially saturated regime.

To see how things go in more detail, let us consider $\langle I(0)I(t) \rangle$. If ξ_1 and ξ_2 are variables associated with $\psi(0)$ and $\psi^*(0)$ and ξ_3 and ξ_4 are the variables associated with $\psi(t)$ and $\psi^*(t)$ then an adequate approximation to M is

$$M \approx \tfrac{1}{2}\Phi^2[1 - \mu(\xi_1 - \xi_2, 0) + 1 - \mu(\xi_3 - \xi_4, 0)] \qquad (11.4.8)$$

in region (a) and

$$\begin{aligned} M \approx &\tfrac{1}{2}\Phi^2[1 - \mu(\xi_1 - \xi_4, 0) + 1 - \mu(\xi_3 - \xi_2, 0)] \\ &+ \Phi^2\{1 - \mu(0, t) - \mu[\tfrac{1}{2}(\xi_1 + \xi_4) - \tfrac{1}{2}(\xi_3 + \xi_2), 0] \\ &+ \mu[\tfrac{1}{2}(\xi_1 + \xi_4) - \tfrac{1}{2}(\xi_3 + \xi_2), t]\} \end{aligned} \qquad (11.4.9)$$

in region (b). The integration over region (a) just gives $\langle I \rangle^2$ so that the integration over region (b) is the interesting one. In the latter region it is convenient to change variables to $\xi = \tfrac{1}{2}(\xi_1 + \xi_4 - \xi_2 - \xi_3)$, $\eta = \tfrac{1}{2}(\xi_1 + \xi_3 - \xi_2 - \xi_4)$, $\zeta = (\xi_1 + \xi_2 - \xi_3 - \xi_4)$ and $\nu = (\xi_1 + \xi_2 + \xi_3 + \xi_4)$. The variable ν does not appear in M and the integration of ν is trivial. It produces a $\delta(\zeta)$ and the ζ integral is then trivial. The final result is the two-fold integral

$$\langle I(t)I(0) \rangle = 1 + (2\pi\Lambda)^{-1} \int_{-\infty}^{\infty} d\eta \, d\xi \, \exp(-i\eta\xi/\Lambda - M')$$

$$(11.4.10)$$

where

$$M' = \Phi^2[1 - \mu(\eta, 0)] + \Phi^2[1 - \mu(0, t) - \mu(\xi, 0) + \mu(\xi, t)]$$
(11.4.11)

is just M restricted to region (b) with $\nu = \zeta = 0$. Specializing to $p = 2$, we have for small η, $1 - \mu(\eta, 0) \approx \frac{1}{2}\eta^2|\ln|\eta||$. Since η will be $\sim \Phi^{-1}$ and hence small, and $|\ln|\eta||$ is slowly varying for small $|\eta|$ we can replace $\eta^2|\ln|\eta||$ by $\eta^2 \ln \Phi$. The η integration then yields

$$\langle I(t)I(0)\rangle = 1 + (\pi\gamma)^{-\frac{1}{2}} \int_{-\infty}^{\infty} d\xi \exp[-\xi^2/\gamma - \bar{M})$$
(11.4.12)

where $\gamma = 2\Phi^2\Lambda^2 \ln \Phi$ and

$$\bar{M} = \Phi^2[1 - \mu(0, t) - \mu(\xi, 0) + \mu(\xi, t)].$$
(11.4.13)

This is just (11.3.15) with an explicit expression, $(\pi\gamma)^{-\frac{1}{2}} \exp(\xi^2\gamma)$, for $P_1(\xi)$ and with $\delta^2\mu(\xi) = \mu(\xi, 0) - \mu(\xi, t)$.

We can now actually compute $\langle I(t)I(0)\rangle$ for partially saturated conditions. In the internal-wave model $\bar{M}$ behaves like $\Phi^2\xi^2t^2a|\ln|\xi||$ for small t and ξ where a is a constant of dimension $(\text{time})^{-2}$ whose order of magnitude is $(\dot{\Phi}/\Phi)^2$. The factor $P_1(\xi)$ will cut off the integration when $\xi \sim \sqrt{\gamma}$ and $|\ln|\xi||$ can be replaced by $\frac{1}{2}|\ln \gamma|$. The net result is then

$$\langle I(t)I(0)\rangle = 1 + (\pi\gamma)^{-\frac{1}{2}} \int_{-\infty}^{\infty} d\xi \exp[-\xi^2(\gamma^{-1} + \frac{1}{2}\Phi^2ta|\ln \gamma|)]$$

$$= 1 + [1 + (t/t_0)^2]^{-\frac{1}{2}}$$
(11.4.14)

where $t_0^{-2} = \frac{1}{2}\gamma\Phi^2a|\ln \gamma| \sim (\Phi\dot{\Phi}\Lambda)^2$. Since $\Phi\Lambda$ is less than unity in the partially saturated regime, t_0 is a time long compared to the time scale $\dot{\Phi}^{-1}$ associated with $D(t)$.

11.5. Correlations in frequency

The correlation between waves with a frequency difference $\sigma - \sigma'$ small compared to the central frequency $\bar{\sigma} = \frac{1}{2}(\sigma + \sigma')$

$$\langle \psi^*(\sigma')\psi(\sigma)\rangle = \exp\left[-\frac{1}{2}\left(\frac{\sigma - \sigma'}{\bar{\sigma}}\right)^2\Phi^2\right]Q$$
(11.5.1)

where

$$Q = \frac{\int_{-\infty}^{\infty} d\xi\, d\xi' \exp\left(-i\frac{\sigma\xi^2 - \sigma'\xi'^2}{2\bar{\sigma}\Lambda} - \frac{1}{2}\Phi^2[1 - \mu(\xi - \xi')]\right)}{\int_{-\infty}^{\infty} d\xi\, d\xi' \exp\left(-i\frac{\sigma\xi^2 - \sigma'\xi'^2}{2\bar{\sigma}\Lambda}\right)}.$$

(11.5.2)

It is understood that Φ and Λ are computed at frequency $\bar{\sigma}$ and in (11.5.2) the approximation

$$\sigma'^2 + \sigma^2 - 2\sigma\sigma'\mu(\xi - \xi') \approx (\sigma - \sigma')^2 + \bar{\sigma}^2[1 - \mu(\xi - \xi')],$$

valid for small $|\sigma - \sigma'|$ and $|\xi - \xi'|$, has been used. Changing variables to $\bar{\xi} = (\sigma\xi - \sigma'\xi')/(\sigma - \sigma')$ and $\eta = \xi - \xi'$ the integral over $\bar{\xi}$ can be effected and cancels between the numerator and denominator. With the further approximation that $\sigma\sigma'/\bar{\sigma}^2 \approx 1$ in the coefficient of $i\eta^2$, the final formula for Q is

$$Q = \frac{\int_{-\infty}^{\infty} d\eta\, \exp\left(-i\frac{\bar{\sigma}\eta^2}{2(\sigma - \sigma')\Lambda} - \frac{1}{2}\Phi^2[1 - \mu(\eta)]\right)}{\int_{-\infty}^{\infty} d\eta\, \exp\left(-i\frac{\bar{\sigma}\eta^2}{2(\sigma - \sigma')\Lambda}\right)}.$$

(11.5.3)

Physically, loss of coherence in frequency is due to fluctuations in the travel time $\bar{\sigma}^{-1}[-\xi_k^2/2\Lambda + \Phi u(\xi_k)]$ along micropaths. The factor

$$\exp\left[-\frac{1}{2}\left(\frac{\sigma - \sigma'}{\bar{\sigma}}\right)^2 \Phi^2\right]$$

in (11.5.1) is the effect of fluctuations in $u(\xi_k)$, while Q is due to variations in $-\xi_k^2/2\Lambda$.

In the saturated regimes higher-order correlations in frequency can be analyzed as follows. For the $2n$th order correlation $\langle \psi^*(\sigma_1')\dots\psi^*(\sigma_n')\psi(\sigma_1)\dots\psi(\sigma_n)\rangle$ there will be the usual $n!$ important regions in the integration volume. For simplicity consider the region where the integration variable ξ_k' associated with $\psi^*(\sigma_k')$ is paired with the variable ξ_k associated with $\psi(\sigma_k)$. In this region an adequate approximation to $M \equiv \frac{1}{2}(\Phi/\bar{\sigma})^2\langle[\sum_{k=1}^{n}\sigma_k'u(\xi_k') - \sigma_k u(\xi_k)]^2\rangle$ is then

$$M \approx \Phi^2 \sum_{k=1}^{n} [1 - \mu(\eta_k)] + M_1$$

(11.5.4)

where

$$M_1 = \tfrac{1}{2}(\Phi/\bar{\sigma})^2 \sum_{k,j=1}^{n} (\sigma'_k - \sigma_k)(\sigma'_j - \sigma_j)u(\bar{\xi}_k - \bar{\xi}_j) \quad (11.5.5)$$

with $\eta_k = \xi_k - \xi'_k$ and $\bar{\xi}_k = \tfrac{1}{2}(\xi_k - \xi'_k)$. Further but different simplifications can be made both in the fully and partially saturated regimes. In the fully saturated regime $u(\bar{\xi}_k - \bar{\xi}_j)$ can be set equal to zero for $k \neq j$ and M_1 becomes $\tfrac{1}{2}\sum_{k=1}^{n}[(\sigma_k - \sigma'_k)\Phi/\bar{\sigma}]^2$. In the partially saturated regime all the $\mu(\bar{\xi}_k - \bar{\xi}_j)$ can be set equal to unity and M_1 is $\tfrac{1}{2}[\sum_{k=1}^{n}(\sigma_k - \sigma'_k)\Phi/\bar{\sigma}]^2$. For either full or partial saturation the $2n$th order correlation is then simply related to the second-order correlation. The statistics in σ for full saturation are Gaussian, while for partial saturation they correspond to a product of a phase factor and a Gaussian. The physical implications of these statistics were discussed in Chapter 8.

CHAPTER 12

PATH INTEGRALS AND PROPAGATION IN SATURATED REGIMES

Propagation through a single upper turning point is easy to analyze because there is an explicit integral representation for ψ. The generalization to propagation through two or more upper turning points is straightforward: ψ can be expressed as a multiple integral that can be studied using the same stationary phase and/or saddle point techniques. Now consider propagation along the sound axis. In this case the scattering is no longer confined to isolated regions and there is no physical phase screen or screens. However, one might try an approximation where all of the scattering is lumped into a screen midway between the source and receiver. Using two screens would be a better approximation, three would be better yet and by letting the number of screens tend to infinity the exact ψ could be recovered. This is just what the Feynman path integral introduced in Chapter 5 does.

It would not be appropriate in this volume to go into the details of the path integral and in any case an attempt to do so is likely to produce a poor imitation of the book of Feynman and Hibbs (1965). We will therefore restrict ourselves to those features of the path integral which are directly relevant to propagation in a random medium. The calculations will be done on a heuristic level and some results will be stated without derivation. For a more precise formulation of path integrals for waves in random media see Dashen (1977).

As the number of 'phase screens' goes to infinity the integration variables become a continuous path and one is actually integrating over a space of functions. We will therefore use a continuum notation. However, for the purposes of this chapter it is best to think of the path integral as an ordinary multiple integral of very large dimension.

This chapter is organized as follows. After reviewing the path integral we show how n-point Gaussian statistics arise in the fully saturated regime. Then a Markov approximation commonly used in optics is introduced. For the internal-wave-induced fluctuations, at least, this Markov approximation is valid for any conceivable propagation geometry and it allows the explicit calculation of $\langle \psi^*(2)\psi(1)\rangle$ in all regimes. Higher-order statistics in the partially saturated regime are treated at the end of the chapter.

Propagation in the unsaturated regimes could also be treated by path-integral techniques. This would however involve considerable calculation and would just reproduce the results of Chapter 10.

12.1. The path integral

The path-integral expression (5.3.6) for ψ can be expressed as

$$\psi = \mathcal{N} \int d(\text{paths}) \exp \left\{ iq_0 S_0(\text{path}) - iq_0 \int_0^R \mu[x, y(x), z(x), t] \, dx \right\}$$

(12.1.1)

where the integration is over all paths $[y(x), z(x)]$ connecting the source to the receiver. The unperturbed phase associated with the path is

$$S_0 = \int_0^R [\tfrac{1}{2}(\partial_x y)^2 + \tfrac{1}{2}(\partial_x z)^2 - U_0(z)] \, dx$$

(12.1.2)

and $\mathcal{N}$ is a normalizing factor to be chosen such that $\psi = 1$ for $\mu = 0$.

It was also shown in Chapter 5 that the ray theory corresponds to a stationary-phase approximation to the path integral. The qualitative conclusions of §11.2 and §11.3 on the phase-screen integral therefore also apply to the path integral. In the fully saturated regime there are many independent stationary-phase points (micropaths) and n-point Gaussian statistics can be expected. As in the phase screen, the stationary-phase points or micropaths are not completely independent in the partially saturated regime. To do quantitative calculations by the stationary-phase method we would need the analog of $P_1(\xi)$ which would be a probability $P_1(\text{path})$ that a path satisfies the ray equation with μ included. A discussion of $P_1(\text{path})$ is available in Dashen (1977). However, in doing our actual calculations we will not take the stationary-phase route. For the phase screen it is simplest, if less physical, to average first and then do the integrals. We will treat the path integral in this way.

To see what kind of calculations need to be done, consider the average of ψ: it is

$$\langle\psi\rangle = \left\langle \mathcal{N}\int d(\text{paths})\exp\left[iq_0 S_0(\text{path}) - iq_0\int_{\text{path}}\mu\,dx\right]\right\rangle$$

$$= \mathcal{N}\int d(\text{paths})\exp\left[iq_0 S_0(\text{path}) - \tfrac{1}{2}q_0^2\left\langle\left(\int\mu\,dx\right)^2\right\rangle\right] \qquad (12.1.3)$$

where we have used the assumption that μ is a Gaussian random variable and the facts that (i) averages and integrations can be interchanged and (ii) for any Gaussian random variable a, $\langle e^{ia}\rangle = e^{-\frac{1}{2}\langle a^2\rangle}$. The second line in (12.1.3) expresses $\langle\psi\rangle$ as a fairly simple looking integral over paths. At this point, however, all the simplicity is in the notation. We have not told the reader how to compute a path integral like this. This will be done later, in § 12.3. As another example of how the notation works consider $\langle\psi^*(2)\psi(1)\rangle$. Multiplying the path integral for $\psi(1)$ by the complex conjugate of the path integral for $\psi(2)$ leads to a 'double path integral' over two paths, one of which connects the source to the observation point (1) and the other connects the source to (2). The result is

$$\langle\psi^*(2)\psi(1)\rangle = \left\langle|\mathcal{N}|^2\int d^2(\text{paths})\exp\left[iq_0 S_0(\text{path }1) - iq_0 S_0(\text{path }2)\right.\right.$$

$$\left.\left. - iq_0\int_{\text{path }1}\mu\,dx + iq_0\int_{\text{path }2}\mu\,dx\right]\right\rangle$$

$$= |\mathcal{N}|^2\int d^2(\text{paths})\exp\left[iq_0 S_0(\text{path }1)\right.$$

$$\left. - iq_0 S_0(\text{path }2) - \tfrac{1}{2}V_{12}\right] \qquad (12.1.4)$$

where

$$V_{12} = q_0^2\left\langle\left(\int_{\text{path }1}\mu\,dx - \int_{\text{path }2}\mu\,dx\right)^2\right\rangle. \qquad (12.1.5)$$

We will also see how to compute this path integral in § 12.3.

12.2. Signal statistics in the fully saturated regime

The n-point Gaussian statistics in the fully saturated regime can be obtained without explicitly doing any path integrals. Consider $\langle I^2\rangle$, it is given by the 'quadruple path integral' over four paths connecting the source to the receiver

$$\langle I^2 \rangle = |\mathcal{N}|^4 \int d^4(\text{path}) \exp\left[-iq_0 \sum_{j=1}^{4} (-1)^j S_0(\text{path } j) - \tfrac{1}{2}M \right] \quad (12.2.1)$$

where

$$M = q_0^2 \left\langle \left(\sum_{j=1}^{4} (-1)^j \int_{\text{path } j} \mu \, dx \right)^2 \right\rangle. \quad (12.2.2)$$

Now in the fully saturated regime where Φ is large, $e^{-\frac{1}{2}M}$ is extremely small except when the paths are pairwise close, that is when path 1 is close to path 2 and path 3 is close to path 4 or when path 1 is close to path 4 and path 2 is close to path 3. Therefore we have an integral much like that in (11.4.4) and we can proceed in analogy with the phase-screen calculation. There are two important regions of path space: (a) where path 1 lies within L_V/Φ of path 2 and path 3 lies within L_V/Φ of path 4 and (b) where path 1 lies within L_V/Φ of path 4 and path 2 lies within L_V/Φ of path 3. The factor $e^{-\frac{1}{2}M}$ does not put any restriction on the distance between pairs of paths: the integration over the separation of pairs is cut off by the oscillating factors $e^{iq_0 S_0}$. In region (a) one can expand the S_0s in the small differences between paths 1 and 2 and between paths 3 and 4. It can then be verified that, just as in the phase screen, the oscillating factors cut off the integral when the distance between pairs is of order $\Phi \Lambda L_V$. There is an identical cutoff in region (b). In the fully saturated regime $\Phi \Lambda$ is large and over most of the integration volume the pairs of paths are separated by more than L_V and therefore are uncorrelated. Thus in both regions (a) and (b) the quadruple path integral can be approximated by a product of double path integrals each of which just gives $\langle I \rangle$. The result is $\langle I^2 \rangle = 2\langle I \rangle^2$.

The calculation of the nth moment of intensity $\langle I^n \rangle$ proceeds in a similar way. There are now $n!$ important regions of path space corresponding to the $n!$ ways of pairing a path from a ψ with another one from a ψ^*. In each of these regions the $2n$-triple path integral can be approximated by the product of n double path integrals with the result that $\langle I^n \rangle = n! \langle I \rangle^n$.

The path-integral method allows the computation of the first corrections to Gaussian statistics (Dashen, 1977). The result is

$$\langle I^n \rangle = n! \langle I \rangle^n [1 + \tfrac{1}{2}n(n-1)\gamma] \quad (12.2.3)$$

where (assuming the Markov approximation described in the next section) γ can be computed explicitly. In a homogeneous isotropic medium there is a fairly simple expression for γ (Dashen, 1977) and it is essentially equal to $(\Phi\Lambda)^{-1}$. For the ocean the calculation of γ involves some rather complicated integrals. The prescription is given in Appendix C. Note that the correction to Gaussian statistics grows with n. The reason is that the oscillating terms in the path integral keep all paths within $\Phi\Lambda L_V$ of each other and if we try to crowd too many pairs of paths into this region they can no longer be considered as uncorrelated. In fact for $n \gtrsim (\Phi\Lambda)^{-1}$ there will be serious departures from Gaussian statistics.

An examination of the expression for γ given in Appendix C shows that γ is also small, $\sim(\ln \Phi)^{-1}$, in the partially saturated regime. This is consistent with our knowledge from the phase screen. In first approximation the intensity statistics in the partially saturated regime are Gaussian but there are relatively large errors of order $(\ln \Phi)^{-1}$.

The method used to relate $\langle I^2 \rangle$ to $\langle I \rangle^2$ can also be used to relate $\langle \psi^*(1)\psi^*(2)\psi(3)\psi(4) \rangle$ to a sum of products of second-order moments. From the point of view of the path integral the only difference is that the paths no longer end at the same points. This is of no particular consequence, however, and as before there are two important regions in path space, and in each of these regions the quadruple path integral can be approximated by a product of two double path integrals. The result is the expected Gaussian rule

$$\langle \psi^*(4)\psi^*(3)\psi(2)\psi(1) \rangle = \langle \psi^*(4)\psi(2) \rangle\langle \psi^*(3)\psi(1) \rangle$$
$$+ \langle \psi^*(4)\psi(1) \rangle\langle \psi^*(3)\psi(2) \rangle.$$

The generalization to an nth moment proceeds in the obvious way.

The first corrections to general n-point Gaussian statistics can be computed (Dashen, 1977). They are of order γ and mainly effect intensity correlations like $\langle I(1)I(2) \rangle$ where they lead to small coherence tails which fall less rapidly than $e^{-D(1-2)}$.

Without actually doing any path integrals we have by now almost derived n-point Gaussian statistics for the fully saturated case. It remains only to show that $\langle (\psi^*)^m (\psi)^n \rangle$ vanishes for $m \neq n$. But this is easy. For $m \neq n$ we cannot pair all the paths associated with a ψ with paths associated with a ψ^*. In all regions of path space the integrand will then be of order $\exp[-\frac{1}{2}(m-n)^2\Phi^2]$ which is vanishingly small for large Φ.

12.3. The Markov approximation

Our goal in this section is to compute $\langle\psi\rangle$ and $\langle\psi^*(1)\psi(2)\rangle$ in the saturated regimes. For simplicity we will first go through the calculation in an idealized model of a two-dimensional homogeneous isotropic medium. Taking the propagation to be in the $x-z$ plane we have

$$\langle\mu(x,z)\mu(x',z')\rangle = \rho((x-x')^2+(z-z')^2)^{\frac{1}{2}}$$

and

$$S_0 = \frac{1}{2}\int_0^R (\partial_x z)^2\,dx.$$

The statistical average which enters into (12.1.3) for ψ is

$$\left\langle\left(\int_{\text{path}}\mu\,dx\right)^2\right\rangle = \int_0^R dx\int_0^R dx'\rho\{(x-x')^2+[z(x)-z(x')]^2\}^{\frac{1}{2}}$$

(12.3.1)

where $z(x)$ is the path. Now $|z(x)-z(x')|/|x-x'|$ is of order $\theta = \partial_x z$ and if the parabolic wave equation is to be valid θ must be small along all important paths. Therefore whenever the parabolic wave equation is valid we can make the approximation

$$\left\langle\left(\int_{\text{path}}\mu\,dx\right)^2\right\rangle \approx \int_0^R dx\int_0^R dx'\,\rho(|x-x'|).$$

(12.3.2)

The path integral in (12.1.3) then reduces to $\exp(-\frac{1}{2}\Phi^2)$ times the path integral for the unperturbed signal and the latter has been normalized to unity. Thus we have $\langle\psi\rangle = \exp(-\frac{1}{2}\Phi^2)$ whenever the parabolic wave equation is valid. The calculation of $\langle\psi^*(2)\psi(1)\rangle$ involves two paths $z_1(x)$ and $z_2(x)$. The V_{12} term in (12.1.4) is explicitly

$$V_{12} = q_0^2\int_0^R dx\int_0^R dx'\{\rho[(x-x')^2+(z_1(x)-z_1(x'))^2]^{\frac{1}{2}}$$
$$+\rho[(x-x')^2+(z_2(x)-z_2(x'))^2]^{\frac{1}{2}}$$
$$-2\rho[(x-x')^2+(z_1(x)-z_2(x'))^2]^{\frac{1}{2}}\}$$

(12.3.3)

where it has been assumed that the time and frequency arguments of $\psi^*(2)$ and $\psi(1)$ are equal. For the same reason as before we can neglect $[z_1(x) - z_1(x')]^2$ relative to $(x - x')^2$ and $[z_2(x) - z_2(x')]^2$ relative to $(x - x')^2$, and in the same spirit we can set $(x - x')^2 + [z_1(x') - z_2(x)]^2 \approx (x - x')^2 + [z_1(\bar{x}) - z_2(\bar{x})]^2$ where $\bar{x} = \frac{1}{2}(x + x')$. Changing variables to $\bar{x}$ and $u = x - x'$ and approximating the limits on u by $\pm\infty$ the u integration just defines the function d and we have

$$V_{12} \approx \int_0^R d[z_1(\bar{x}) - z_2(\bar{x})] \, d\bar{x}. \qquad (12.3.4)$$

With this approximation for V_{12} the path integral in (12.1.4) can be done exactly by changing variables to paths $v(x) = z_1(x) - z_2(x)$ and $w(x) = \frac{1}{2}[z_1(x) + z_2(x)]$. In these variables the path integral is explicitly

$$\langle \psi^*(2)\psi(1) \rangle = |\mathcal{N}|^2 \int d^2(\text{paths})$$

$$\cdot \exp\left\{ iq_0 \int_0^R \partial_x v(x) \, \partial_x w(x) \, dx - \frac{1}{2} \int_0^R d[v(x)] \, dx \right\}$$

$$(12.3.5)$$

and without loss of generality the endpoint conditions on the paths v and w can be taken to be $v(0) = w(0) = w(R) = 0$ and $v(R) = \Delta z$, the separation between receivers. Note that the path w appears only as a linear factor in the argument of the exponential and in analogy with the formula

$$\int_{-\infty}^{\infty} d\alpha \; e^{i\alpha\beta} = 2\pi\delta(\beta) \qquad (12.3.6)$$

one would expect the integral over the paths $w(x)$ to produce a product of δ-functions. Integrating by parts so that $\int_0^R \partial_x v \, \partial_x w \, dx$ becomes $-\int_0^R \partial_x^2 v w \, dx$ and writing out the path integral in its finite form as an ordinary multidimensional integral one finds that this is exactly what happens. The integration over the now discrete ws yields a product of δ-functions which forces v to satisfy the finite difference analog of $\partial_x^2 v(x) = 0$ everywhere. Knowing that v will be restricted by δ-functions we can replace the v in $d(v)$ by the solution to $\partial_x^2 v = 0$ with boundary conditions $v(0) = 0$ and $v(R) = \Delta z$, namely $v(x) = \Delta z x / R$. Noting that $\int_0^R d(\Delta z x / R) \, dx = D(\Delta z)$ we then have a product of $\exp\left[-\frac{1}{2}D(\Delta z)\right]$ and the path integral for the unperturbed intensity. The result is $\langle\psi^*(2)\psi(1)\rangle = e^{-\frac{1}{2}D(1-2)}$ which has actually been derived only for equal times but the generalization to different times is straightforward.

For a homogeneous isotropic medium the validity of the above approximations can be checked by explicitly computing the first corrections (Dashen, 1977). The corrections are fractionally of order of the mean-square scattering angle and are therefore small whenever the parabolic wave equation is valid.

In optics (Tatarskii, 1971), the approximation in (12.3.4) is called a Markov approximation. It shows up in a set of partial differential equations which are equivalent to the path integral in (12.1.4). The term Markov is used because it basically says that the scattering at a given range point is independent of the scattering at other range points, i.e. it is a random process with 'no memory'. In an isotropic medium with small μ this is equivalent to saying that the rms scattering angle is small. As long as the wave is going in essentially one direction there cannot be any significant correlations with distant scatterings. There is a potential problem with a Markov approximation in a highly anisotropic medium. When the scattering by a given inhomogeneity is highly dependent on the angle of incidence there can be important correlations with distant scatterings which have changed the angle of incidence. Beran, McCoy and Adams (1975) have studied a limiting case of this situation. Fortunately, even though the ocean is anisotropic we do not have this problem. The background sound channel saves the day. For the flat pancake-like inhomogeneities that exist in the ocean, the scattering is strongly dependent on angle only when the angle of incidence is less than the aspect ratio of the inhomogeneities. The aspect ratio for internal-wave-induced fluctuations is of order ω_i/n and it is only very close to turning points that rays can have a slope less than ω_i/n. Distant scatterings may change the locations of turning points but they will not affect the way that the ray slope varies as the turning point is traversed. Since the average scattering around a turning point is only weakly dependent on its location there is no important 'memory' of previous scatterings. In the ocean, a Markov approximation will be valid out to ranges where variations in the depths of turning points become important. For internal waves one estimates that this will happen only at ranges $\sim 10^5$ km which are of no practical interest.

In the Markov approximation we can always set $q_0^2 \langle (\int_{\text{path}} \mu \, dx)^2 \rangle = \Phi^2$ in (12.1.3) obtaining immediately $\langle \psi \rangle = e^{-\frac{1}{2}\Phi^2}$. The analog of (12.3.4) for the real ocean is

$$V_{12} = \int_0^R d[y_1(x) - y_2(x), z_1(x) - z_2(x), t_1 - t_2; x] \, dx,$$

$$(12.3.7)$$

and with this approximation for V_{12} the path integral in (12.1.4) can be done in almost exactly the same way as it was for the two-dimensional isotropic model. Define new paths by

$$z_1 = z_0 + w + \tfrac{1}{2}v,$$

$$z_2 = z_0 + w - \tfrac{1}{2}v,$$

$$y_1 = y_0 + u + \tfrac{1}{2}s,$$

and

$$y_2 = y_0 + u - \tfrac{1}{2}s,$$

where $[z_0(x), y_0(x)]$ is the unperturbed ray. Expanding the unperturbed phase $S_0(\text{path } 1) - S_0(\text{path } 2)$ in powers of w, v, u and s one finds that there are no linear terms (because (z_0, y_0) satisfies the ray equation) and that the quadratic terms are of the form wv and su. Neglecting cubic and higher terms we then have a path integral where w and u appear only as linear factors in the argument of the exponential. Integrating over w and u will then produce a product of δ-functions which force v and s to satisfy the equations for small deviations from a ray. The integral over d then becomes D as before and the remaining path integral just gives the unperturbed intensity which is normalized to unity. The result is

$$\langle \psi^*(2)\psi(1) \rangle = \exp\left[-\tfrac{1}{2}D(1-2)\right].$$

To complete the specification of the second moments we need to compute $\langle \psi^*(\sigma_2)\psi(\sigma_1) \rangle$. We will use the Markov approximation from the beginning and at the expense of a tiny error of order $(L_V/L_H)^2$ we will also ignore the y components of the paths. Up to a normalization we then have, for $|\sigma_1 - \sigma_2| \ll \bar{\sigma}$ and $\bar{\sigma} = \frac{1}{2}(\sigma_1 + \sigma_2)$

$$\langle \psi^*(\sigma_2)\psi(\sigma_1) \rangle = \exp\left[-\frac{1}{2}\left(\frac{\sigma_1 - \sigma_2}{\bar{\sigma}}\right)^2 \Phi^2 \right] \int d^2(\text{paths})$$

$$\cdot \exp\left\{ i\frac{\sigma_1}{C_0} S_0(\text{path 1}) \right.$$

$$\left. -i\frac{\sigma_2}{C_0} S_0(\text{path 2}) - \frac{1}{2}\int_0^R d[z_1(x) - z_2(x), x]\,dx \right\}$$

(12.3.8)

where Φ^2 and d are evaluated at frequency $\bar{\sigma}$. Changing variables to u and v defined by $z_1 = z_0 + u - \sigma_2 v/(\sigma_1 - \sigma_2)$ and $z_2 = z_0 + u - \sigma_1 v/(\sigma_1 - \sigma_2)$ where z_0 is the unperturbed ray, one expands the unperturbed phases $S_0(\text{path 1}) - S_0(\text{path 2})$ in powers of u and v keeping only quadratic terms (again there are no linear terms). The argument of the exponential is then the sum of a term involving u and another involving v. The integral over u can then be done separately and since it involves only unperturbed quantities it contributes only to the normalization. The result of this sequence of approximations and manipulations is

$$\frac{\langle \psi^*(\sigma_2)\psi(\sigma_1) \rangle}{\psi_0^*(\sigma_2)\psi_0(\sigma_1)} = \exp\left[-\frac{1}{2}\left(\frac{\sigma_1 - \sigma_2}{\bar{\sigma}}\right)^2 \Phi^2 \right] Q \qquad (12.3.9)$$

with

$$Q = \frac{\displaystyle\int d(\text{paths}) \exp\left[-\frac{i\bar{\sigma}^2}{2C_0(\sigma_1 - \sigma_2)} \int_0^R \{[\partial_x v(x)]^2 - U_0''(x)v^2(x)\}\,dx - \int_0^R d[v(x), x]\,dx \right]}{\displaystyle\int d(\text{paths}) \exp\left[-\frac{i\bar{\sigma}^2}{2C_0(\sigma_1 - \sigma_2)} \int_0^R \{[\partial_x v(x)]^2 - U_0''(x)v^2(x)\}\,dx \right]}$$

(12.3.10)

where $U_0''(x) = \partial_z^2 U_0[z_0(x)]$ and the correct normalization has been obtained by dividing by the same path integral for $\mu = 0$. For large Φ, the only paths which make a significant contribution are those for which $|v| \lesssim L_V/\Phi$. Assuming for the moment that $d(z)$ can be expanded around $z = 0$, $\int_0^R d[v(x), x] \, dx$ can then be approximated by $\frac{1}{2} \int_0^R v^2(x) \partial_z^2 d(0, x) \, dx$. This produces a path integral whose integrand is the exponential of a quadratic function of the path v. If the path integral were written out in its finite form as a multi-dimensional integral then we would have an ordinary integral whose integrand is the exponential of a quadratic form in the integration variables. Such an integral is equal to the determinant of the quadratic form and in particular Q would be the ratio of two determinants. In the limit in which the path integral becomes an integration over functions this ratio of determinants becomes a ratio of functional determinants (loosely speaking, Fredholm determinants). This ratio of functional determinants can be calculated either by computing ratios of eigenvalues or by solving an initial value problem. The two methods are explained in Appendix B.

For the internal-wave spectrum $d(z)$ behaves like $z^2 \ln z$ at small z and strictly speaking the expansion which leads to a quadratic form is not valid. However, as in the phase screen calculations of Chapter 11 this problem can be avoided by interpreting $\ln v$ as $\ln \Phi$.

The specification of $\langle \psi^*(2)\psi(1) \rangle$ in the saturated regimes is now complete.

12.4. The partially saturated regime

It has already been remarked in § 12.2 that the intensity statistics are Gaussian in the partially saturated regime with errors of order $(\ln \Phi)^{-1}$. The same is true of the general n-point statistics at equal times and frequencies. This has been verified by an explicit calculation of the first corrections.

At unequal frequencies the situation is exactly the same as in the phase-screen model. In the partially saturated regime ψ is a product of a Gaussian phase factor and a complex Gaussian random variable whose variance is given by Q. This result can be obtained by manipulating path integrals in exactly the same way as the ordinary integrals were manipulated in § 11.5.

At unequal times there is again a complete analogy with the phase screen. Manipulating the path integrals in exactly the same way as the phase-screen integral, one finds the analog of (11.4.10)

$\langle I(t)I(0)\rangle - 1$

$$= \int d^2(\text{paths})$$

$$\cdot \exp\left\{-iq_0 \int_0^R [\partial_x\xi(x)\,\partial_x\eta(x) - U_0''(x)\eta(x)\xi(x)]\,dx - M'\right\}$$

$$= \frac{}{\int d^2(\text{paths}) \exp\left\{-iq_0 \int_0^R [\partial_x\xi(x)\,\partial_x\eta(x) - U_0''(x)\eta(x)\xi(x)]\,dx\right\}}$$

$$\tag{12.4.1}$$

where $U_0''(x)$ has the same meaning as before, $\eta(x)$ and $\xi(x)$ are the z-components of two paths and now

$$M' = \int_0^R d[\eta(x), 0; x]\,dx$$

$$+ \int_0^R \{d(0, t; x) + d[\xi(x), 0; x] - d[\xi(x), t; x]\}\,dx. \tag{12.4.2}$$

To evaluate $\langle I(t)I(0)\rangle$ one expands M' in powers of η, ξ and t keeping only the leading terms which are η^2 and $\xi^2 t^2$ and if necessary interpreting $|\ln |\eta||$ as $\ln \Phi$ and $|\ln |\xi||$ as $\frac{1}{2}\ln (2\Phi^2\Lambda^2 \ln \Phi)$ as in § 11.4. The integrand is then the exponential of a quadratic function of the paths, and $\langle I(t)I(0)\rangle - 1$ is a ratio of determinants. Methods for computing these determinants are given in Appendix A.

We now have everything except (9.4.15) whose derivation will now be outlined. One has to study third and higher moments of the intensity. As is always the case in the saturated regimes the main contribution comes from regions where paths are pairwise closer than L_V/Φ. To obtain (9.4.15) one treats each region separately and in the following manner. Let η_k be the difference between paired paths and ξ_k be the centroids of the pairs where k runs over the path pairs. Then in correlations $\langle\int_{\text{path }i} \mu\,dx \int_{\text{path }j} \mu\,dx\rangle$ between paths that are not paired set the η_ks equal to zero. Having done this expand the argument of the exponential in powers of η_ks, ξ_ks and t. The leading terms are of the form η_k^2 and $\xi_k^2 t^2$. Then by choosing new variables which are linear combinations of the old paths it is possible to factorize the path integral into a product of path integrals, one of which is exactly the same path integral that appears in the calculation of $\langle I(t)I(0)\rangle$. All the other path integrals in the product are trivial to evaluate and contribute only to the normalization. In (9.4.15) the sum over permutations is a sum over the dominant regions of path space and the function K is the result of having transformed the integral over a given region into a form of the basic integral for $\langle I(t)I(0)\rangle$.

CHAPTER 13

THE TRANSPORT EQUATION
IN SOUND SCATTERING

A description of wave propagation through a randomly varying medium may often, and conveniently, be given using a transport equation. Familiar examples include electromagnetic radiative transport theory (see Chandrasekhar, 1950; or Bond *et al.*, 1965) and the Boltzmann equation for non-equilibrium gaseous phenomena (wave propagation, if the Schrödinger equation is considered as basic). This description of scattering is restricted to evaluation of the radiant energy flux (defined in § 13.1), and is equivalent in the Fourier-transform sense to determining the propagation of the second moment $\langle \psi^*(2)\psi(1) \rangle$. Hence the transport method yields results that are closely related to results obtained from the propagation of the second moment in the parabolic-equation (small scattering angle) approximation. The path-integral technique, described in Chapters 8, 11 and 12, gives results for all the moments, and hence also for the second moment. However, effects which are conveniently expressed in terms of scattering angle (such as scattering loss outside an acceptance cone) are best treated from the point-of-view of the transport equation.

In § 13.2 the transport equation is exhibited, its relation to the second moment is described, and the scattering cross-section is related to the spectrum of sound-speed fluctuations. In § 13.3 the small-angle approximation is made, and the resulting transport equation is related to the well-known differential equation for the propagation of the second moment, and to the solution to that differential equation provided by the path-integral technique. The diffusion approximation is then made and related to the Gaussian approximation for the second moment. In § 13.4 the particular case of scattering through internal waves, in which the sound-speed fluctuations are larger than an acoustic wavelength, is treated. The probability distribution of scattering angle is calculated, and the resulting absorption coefficient is evaluated. In § 13.5 large-angle scattering from inhomogeneities that are smaller than a wavelength is treated; the scattering loss is calculated in terms of the mean-square temperature gradient.

13.1. The energy flux

To describe sound transmission through a random internal-wave field in the transport approximation, an energy intensity $I_\sigma(\mathbf{x}, \hat{\mathbf{q}})$ is introduced. This is so defined that

$$I_\sigma(\mathbf{x}, \hat{\mathbf{q}}) \, d\sigma \, d\Omega \hat{\mathbf{q}} \qquad (13.1.1)$$

represents the flow of acoustic energy per unit area at the position $\mathbf{x}$, having angular frequency σ within the interval $d\sigma$, and propagating in the direction $\hat{\mathbf{q}}$ within the solid angle $d\Omega \hat{\mathbf{q}}$. When it is not desired to specify the frequency, the intensity

$$I(\mathbf{x}, \hat{\mathbf{q}}) = \int_0^\infty I_\sigma(\mathbf{x}, \hat{\mathbf{q}}) \, d\sigma \qquad (13.1.2)$$

is used. If the energy flux is confined to a narrow angular cone, the flux per unit area

$$F(\mathbf{x}) = \int I(\mathbf{x}, \hat{\mathbf{q}}) \, d\Omega \hat{\mathbf{q}} \qquad (13.1.3)$$

may be adequate to describe the acoustic field. The acoustic intensity I_σ is considered to represent a suitable ensemble average over the fluctuations of the random scattering medium. To measure I_σ observation times must therefore be large compared with important time scales for the random variation of scattering medium.

As we shall see in this chapter, the transport equation to determine the intensity is much simpler to treat than is the full wave equation. This advantage of simplicity is offset, however, by certain limitations. First, complete phase information is lacking in the transport description of wave propagation. Second, the transport equation in general represents an approximation to the wave equation and can be used only when this approximation is adequate. Third, fluctuation phenomena are not fully described by the ensemble averaged intensity I_σ, because moments higher than second-order are not treated.

A variety of derivations of transport equations from wave equations have been published. These differ in details and make approximations suitable for the specific phenomena studied, but generally tend to be rather similar. One of the first such derivations was given by Foldy (1945). He showed that for scalar waves scattered by uncorrelated (in position) point scatterers, the transport equation is exact if the number of scatterers is large compared to unity. Subsequent derivations have studied correlated scatterers. To obtain a valid transport equation in such cases it is necessary that the scattering be sufficiently weak that I_σ changes little over a correlation distance, L_{corr}, of the scatterers. Snider (1960) and Watson (1960) gave derivations of particle transport in a gas-like medium, starting from Schrödinger's wave equation. Derivations of the radiative transfer equation, starting from Maxwell's equations, have been given by Stott (1968) and Watson (1969). Hasselmann (1962, 1963a, b) has obtained an equation to describe the transport of ocean surface wave energy.

13.2. The transport equation for acoustic intensity

To describe sound propagation in the fluctuating internal-wave field, we first rewrite (5.1.11) for the acoustic pressure $p(\mathbf{x}, t)$ as

$$\{\nabla^2 - [C_0(1 + U_0(z))]^{-2} \partial_{tt}\}p = 2\mu C_0^{-2} \partial_{tt} p, \qquad (13.2.1)$$

using the notation (6.1.1) for the sound speed. In the absence of inhomogeneities we would have $\mu = 0$. We shall assume in this chapter that when $\mu = 0$ the eikonal approximation (Chapter 5) may be used to determine p. In particular, we suppose that sound propagates along ray paths determined from the equation

$$\frac{d\hat{\mathbf{q}}}{ds} = (\nabla - \hat{\mathbf{q}}\hat{\mathbf{q}} \cdot \nabla) \ln (1 + U_0). \qquad (13.2.2)$$

Here $\hat{\mathbf{q}}$ is a unit vector tangent to a given ray path and ds is an element of length along that path.

Equation (13.2.1) is linear in p, so we may take p to be a complex quantity having elementary 'local' solutions of the form $p \sim \exp[i(\mathbf{q} \cdot \mathbf{x} - \sigma t)]$, where $\sigma \approx qC_0(1 + U_0)$. The appropriate definition of acoustic intensity for this phase convention has been given by Wigner (1932) as

$$I_\sigma(\mathbf{x}, \hat{\mathbf{q}}) = \left[\frac{(2\pi)^{-4}}{2\rho_0 C_0}\right] \int d^3r \, e^{-i\mathbf{q}\cdot\mathbf{r}} \int d\tau \, e^{i\sigma\tau}$$

$$\cdot \langle p^*(\mathbf{x} - \mathbf{r}/2, t - \tau/2) p(\mathbf{x} + \mathbf{r}/2, t + \tau/2) \rangle,$$

(13.2.3)

where $\langle \cdots \rangle$ represents the ensemble average, indicated earlier, over fluctuations of the scattering medium. If we integrate (13.2.3) over frequency, as in (13.1.2), we obtain an expression in terms of the quantity ψ introduced in (8.1.3) and (8.1.4);

$$I(\mathbf{x}, \mathbf{q}) = \left(\frac{(2\pi)^{-3}}{4\rho_0 C_0}\right) \int d^3r \, e^{-i\mathbf{q}\cdot\mathbf{r}} |f_0|^2 \langle \psi^*(\mathbf{x} - \mathbf{r}/2, t) \psi(\mathbf{x} + \mathbf{r}/2, t) \rangle.$$

(13.2.4)

The description of propagation of the intensity is thus seen to be equivalent to that for the second moment $\langle \psi^*(2)\psi(1) \rangle$.

The derivation of the transport equation for the intensity starting from (13.2.1) is relatively straightforward, but tedious, and will not be reproduced here.[†] When Doppler shifting of the frequency σ may be neglected, the transport equation is of the form

$$\left(\frac{d}{ds} + \alpha\right) I_\sigma(\mathbf{x}, \hat{\mathbf{q}}) = \int d\Omega_{\hat{\mathbf{q}}'} \, \Sigma(\hat{\mathbf{q}}, \hat{\mathbf{q}}') I_\sigma(\mathbf{x}, \hat{\mathbf{q}}').$$

(13.2.5)

Here the derivative is evaluated at $\mathbf{x}$ and in the direction $\hat{\mathbf{q}}$ of the tangent to a ray path passing through $\mathbf{x}$, defined in (13.2.2). The expression $\Sigma(\hat{\mathbf{q}}, \hat{\mathbf{q}}')$ represents the cross-section per unit volume for scattering of a ray from the direction $\hat{\mathbf{q}}'$ to the direction $\hat{\mathbf{q}}$. The quantity α in (13.2.5) is expressed as the sum of four terms:[‡]

$$\alpha = \alpha_d + \alpha_s + \alpha_r + \alpha_t.$$

(13.2.6)

Here α_d represents absorption due to molecular dissipative processes, including viscosity, and α_s represents absorption due to scattering by solid matter, including plankton and fish (Weston, 1967; Proni and Apel, 1975). The term α_r describes variations in

---

[†] The reader requiring a derivation should refer to one of the derivations mentioned in § 13.1. The extension to curved ray paths was given by Lau and Watson (1970).
[‡] Since we are not giving a derivation of (13.2.5), we omitted from (13.2.1) the terms corresponding to α_d and α_s.

intensity due to refraction and has the form

$$\alpha_r = -\frac{d}{ds} \ln(1 + U_0)^2.$$ (13.2.7)

Finally, α_t describes the decay of intensity on a given ray path due to scattering by the internal-wave fluctuations:

$$\alpha_t = \int d\Omega q \, \Sigma(\hat{q}, \hat{q}').$$ (13.2.8)

The right-hand side of (13.2.5) tends to compensate for scattering loss by describing scattering of acoustic energy into that ray path with tangent $\hat{q}$.

Several conditions must be met (Watson, 1969) if (13.2.5) is to be an adequate approximation and replacement for (13.2.1):

(1) We require that $\alpha/q \ll 1$. This implies little absorption in one acoustic wavelength and is well satisfied for our intended applications.

(2) The eikonal approximation, as discussed in Chapter 5, must be valid in the absence of internal-wave fluctuations.

(3) The intensity function $I_\sigma(\mathbf{x}, \hat{q})$ must not change significantly over either of the correlation distances L_H or L_V (6.1.11). This condition is fundamental for the validity of (13.2.5). Were this condition violated, rays coming from different directions $\hat{q}'$ and $\hat{q}''$ could coherently interfere on being scattered into the direction $\hat{q}$ which is an effect evidently not included in (13.2.5).

The cross-section $\Sigma(\hat{q}, \hat{q}')$ is easily calculated in either the Born or the eikonal approximation.[†] From (10.1.5) we obtain the scattering amplitude in Born approximation as

$$f(\hat{q}, \hat{q}') = -\left(\frac{q^2}{2\pi C_0}\right) \int d^3x \, e^{iq(\hat{q}'-\hat{q})\cdot\mathbf{x}}[C_0\mu(\mathbf{x})].$$ (13.2.9)

The cross-section Σ is then

$$\Sigma(\hat{q}, \hat{q}')$$

$$= \Sigma(\hat{q}', \hat{q}) = \frac{\langle|f(\hat{q}, \hat{q}')|^2\rangle}{\text{volume}}$$

$$= \left(\frac{q^2}{2\pi}\right)^2 \int d^3x \, d^3y \, e^{-i\mathbf{k}\cdot(\mathbf{x}-\mathbf{y})}\langle\mu(\mathbf{x})\mu(\mathbf{y})\rangle \, [\text{volume}]^{-1},$$ (13.2.10)

[†] It may be shown that for our applications these two methods of evaluating Σ lead to equivalent results.

where

$$\mathbf{k} \equiv q(\hat{\mathbf{q}} - \hat{\mathbf{q}}').$$ (13.2.11)

The 'volume' in the denominator above is that over which the integration is done. It should be comparable to the 'correlation volume' $\pi L_H^2 L_V$. The autocorrelation appearing in (13.2.10) may be expressed from (3.6.4):

$$\langle \mu(\mathbf{x} + \mathbf{x}'/2)\mu(\mathbf{x} - \mathbf{x}'/2) \rangle = \int e^{i\mathbf{k} \cdot \mathbf{x}'} F(\mathbf{k}, z) \, d^3 k,$$ (13.2.12)

where F represents the power spectrum (3.6.7) of sound-speed fluctuations. We thus find that

$$\Sigma(\hat{\mathbf{q}}, \hat{\mathbf{q}}') = 2\pi q^4 F(\mathbf{k}, z).$$ (13.2.13)

13.3. The diffusion approximation

In this and the following sections we specialize our discussion to ray paths which are approximately horizontal. This is evidently the case of most practical significance when studying scattering by fluctuations induced in the sound speed by internal waves. We shall also assume that the acoustic wavelength is small compared with that of the internal waves responsible for the scattering. This implies that individual scatterings will be through small angles and permits us to simplify (13.2.5) with the diffusion approximation.

To do this, it is convenient to first so orient our coordinate system that the x-axis is locally very nearly parallel to the ray paths of interest. We then introduce the vector θ having components θ_y and θ_z, where θ_z is the vertical component of $\mathbf{q}/q$ and θ_y is that component of $\mathbf{q}/q$ parallel to the y-axis. We may assume I_σ and Σ to be functions of θ and set

$$d\Omega \hat{\mathbf{q}} = d^2 \theta$$ (13.3.1)

in (13.2.5). This equation now reads

$$\left(\frac{d}{ds} + \alpha\right) I_\sigma(\mathbf{x}, \theta) = \int d^2 \theta' \, \Sigma(\theta, \theta') I_\sigma(\mathbf{x}, \theta').$$ (13.3.2)

We have seen in (13.2.4) that the Fourier transform of I_σ (let us denote it by $\tilde{I}_\sigma$) is proportional to the second-moment function. If

we write $\mathbf{k} \equiv q\boldsymbol{\theta}$, $d/ds = \partial/\partial x + \mathbf{k} \cdot \nabla_x$,
and

$$I_\sigma(\mathbf{x}, \boldsymbol{\theta}) = (2\pi)^{-2} \int d^3r \, e^{i\mathbf{k}\cdot\mathbf{r}} \tilde{I}_\sigma(\mathbf{x}, \mathbf{r}),$$

$$\Sigma(\boldsymbol{\theta}, \boldsymbol{\theta}') = (2\pi)^{-2} \int d^3r \, e^{i(\mathbf{k}'-\mathbf{k})\cdot\mathbf{r}} \Sigma(\mathbf{r}),$$

equation (13.3.2) takes the form

$$\left(\frac{\partial}{\partial x} - i\nabla_r \cdot \nabla_x + \alpha\right)\tilde{I}_\sigma(\mathbf{x}, \mathbf{r}) = \tilde{\Sigma}(\mathbf{r})\tilde{I}_\sigma(\mathbf{x}, \mathbf{r}). \tag{13.3.3}$$

This differential equation for the propagation of the second moment has been studied extensively in the literature, and solutions for many special cases have been established. Reviews of the subject appear in Barabanenkov *et al.* (1970), Prokhorov *et al.* (1975), and Beran (1975). The path-integral technique provides an *explicit, formal solution* to (13.3.3). This solution is given as (12.3.5).

A further approximation, known as the *diffusion approximation*, may be made at this point, provided that the spectrum of sound-speed fluctuations is appropriately behaved. Suppose we may keep only the first three terms of the Taylor expansion

$$I_\sigma(\mathbf{x}, \boldsymbol{\theta}') = I_\sigma(\mathbf{x}, \boldsymbol{\theta}) + \sum_{i=1}^{2} \Delta\theta_i \frac{\partial I_\sigma(\mathbf{x}, \boldsymbol{\theta})}{\partial \theta_i}$$

$$+ \frac{1}{2} \sum_{i,j=1}^{2} \Delta\theta_i \, \Delta\theta_j \frac{\partial^2 I_\sigma(\mathbf{x}, \boldsymbol{\theta})}{\partial \theta_i \, \partial \theta_j} + \cdots,$$

where $\Delta\boldsymbol{\theta} \equiv \boldsymbol{\theta}' - \boldsymbol{\theta}$. Terminating this series is equivalent to assuming that the second moment $\langle \psi^*(2)\psi(1) \rangle$ is a Gaussian as a function of transverse spatial separations. Recalling (8.5.1), this amounts to approximating the phase-structure function D by a quadratic, see for example (7.5.14). Substitution of the first three terms of this expansion into (13.3.2) gives us the diffusion equation

$$\left(\frac{d}{ds} + \alpha_0\right)I_\sigma(\mathbf{x}, \boldsymbol{\theta}) = \kappa_z \frac{\partial^2 I_\sigma}{\partial \theta_z^2} + \kappa_y \frac{\partial^2 I_\sigma}{\partial \theta_y^2}. \tag{13.3.4}$$

Here we have used the relations

$$\int d^2\theta' \, \Sigma = \alpha_t,$$

$$\int d^2\theta' \, \Delta\boldsymbol{\theta} \, \Sigma = 0,$$

$$\int d^2\theta' \, \Delta\theta_y \, \Delta\theta_z \, \Sigma = 0,$$

and the definitions

$$\kappa_y \equiv 2^{-1} \int d^2\theta' \, \Delta\theta_y^2 \Sigma,$$

$$\kappa_z \equiv 2^{-1} \int d^2\theta' \, \Delta\theta_z^2 \Sigma, \tag{13.3.5}$$

$$\alpha_0 = \alpha - \alpha_t.$$

We see from (13.3.4) that the total flux

$$F = \int I_\sigma \, d^2\theta$$

satisfies the equation

$$\left(\frac{d}{ds} + \alpha_0\right)F = 0. \tag{13.3.6}$$

The implication of this equation is that the small-angle scatterings resulting from internal waves do not contribute to the decay of the flux F.

Similarly, we can define mean-square scattering angles in the vertical and horizontal directions as

$$F\overline{\theta_y^2} \equiv \int d^2\theta \, \theta_y^2 I_\sigma,$$

$$F\overline{\theta_z^2} \equiv \int d^2\theta \, \theta_z^2 I_\sigma. \tag{13.3.7}$$

From (13.3.4) and these definitions we obtain the relations

$$\frac{d}{ds} \overline{\theta_y^2} = 2\kappa_y,$$

$$\frac{d}{ds} \overline{\theta_z^2} = 2\kappa_z.$$

These may be integrated along a ray path to give

$$\overline{\theta_y^2} = 2 \int \kappa_y \, ds,$$

$$\overline{\theta_z^2} = 2 \int \kappa_z \, ds. \tag{13.3.8}$$

Since κ_y and κ_z vary with the depth $(-z)$, we must in general insert the value $z = z(s)$ along the ray path when performing the above integrations.

Since θ_y and θ_z represent angular deflections from a ray path, it is only the internal-wave scattering which contributes to these.

13.4. Scattering from internal waves

First, we evaluate the diffusion constants κ_y and κ_z using the internal-wave spectrum (3.6.7). On writing $k = q\theta_y$, $k_v = q\theta_z$, we have, from (13.3.5) and (13.2.12)

$$\kappa_y = \pi \int_0^{k_{v\,\mathrm{max}}} dk_v \int_0^{k_v} k^2 F(\mathbf{k}, z) \, dk.$$

We are required here to assign an upper limit to the k_v-integration. Following Garrett and Munk (1975) we take $k_{v\,\mathrm{max}} = 870\, k_v^*$. Evaluation then gives

$$\kappa_y \approx (2/\pi)\langle\mu^2\rangle(\omega_i/n_0)(j_*/B) \ln(n/\omega_i) \ln(870). \tag{13.4.1}$$

Similarly, we find that

$$\kappa_z = \pi \int_0^{k_{v\,\mathrm{max}}} dk_v \int_0^{k_v} k_v^2 F(\mathbf{k}, z) \, dk$$

$$\approx (1/\pi)\langle\mu^2\rangle(n/n_0)(n/\omega_i)(j_*/B) \ln(870). \tag{13.4.2}$$

To calculate the absorption coefficient α_t (13.2.8), we must assign a lower limit to the k_v-integration. This presumably lies between the (ocean depth)$^{-1}$ and k_v^*. The lower limit appears only in the argument of a logarithm and we take this to be k_v^*. Then, we write

$$\alpha_t = 2\pi q^2 \int_{k_v^*}^{\infty} dk_v \int_0^{k_v} F(\mathbf{k}, z) \, dk$$

$$\approx (\pi^{-3})\langle\mu^2\rangle(n_0/\omega_i)(B/j_*)(\ln 2)q^2. \tag{13.4.3}$$

On setting $j_* = 3$, $B = 1200$ m, and scaling to the depth $z = -B$, we find that [†]

$$\kappa_y = 5.7 + 10^{-12}[n(z)/n(-B)]^3 \text{ m}^{-1},$$

$$\kappa_z = 6.0 \times 10^{-10}[n(z)/n(-B)]^5 \text{ m}^{-1},$$

$$\alpha_t = 1.3 \times 10^{-4} f^2 [n(z)/n(-B)]^3 \text{ m}^{-1} \qquad (13.4.4)$$

$$= 0.56 f^2 [n(z)/n(-B)]^3 \text{ dB km}^{-1}.$$

Here f is the acoustic frequency in kHz. The resulting rms scattering angles for a ray on the sound-channel axis are obtained from (13.3.8) and (13.4.4) as

$$(\overline{\theta_y^2})^{\frac{1}{2}} = 1.1 \times 10^{-4} R^{\frac{1}{2}},$$

$$(\overline{\theta_z^2})^{\frac{1}{2}} = 1.1 \times 10^{-3} R^{\frac{1}{2}}, \qquad (13.4.5)$$

where R is the path length in km.

As would be expected to result for the relatively smaller scale of vertical than horizontal wavelengths, the vertical scattering is substantially greater than that in the horizontal direction. We see, however, that the actual scattering angles tend to be small, as was assumed in the expansion leading to (13.3.4). The probability distribution P of vertical scattering angles θ_z may be obtained from the expression

$$P(\theta_z/\theta_*) = \text{constant} \times \int d\theta_y \Sigma. \qquad (13.4.6)$$

This may be expressed as a function of the scaled variable θ/θ_*, where (on the channel axis)

$$\theta_* = \frac{6.8 \times 10^{-4}}{f} \text{ radians} = \frac{4.0 \times 10^{-2}}{f} \text{ degrees.} \qquad (13.4.7)$$

The probability P is shown as a function of θ_z/θ_* in Fig. 13.1. The expected predominance of very small scattering angles is evident for frequencies in the 50–100 Hz range, or higher.

It was noted in connection with (13.3.6) that α_t does not represent true absorption, but just scattering away from a given ray. To better interpret this quantity, we might define an 'acceptance cone', corresponding to $\theta_* < \theta_0$. If we interpret as 'lost' any energy for

[†] We have here evaluated $\ln (n/\omega_i)$ in (13.4.1) as $\ln [n(-B)/\omega_i]$.

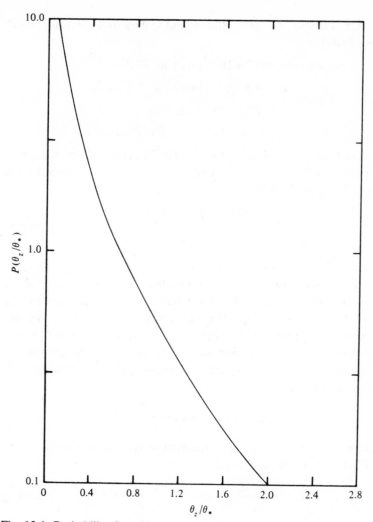

Fig. 13.1. Probability (in arbitrary units) of scattering through a vertical angle θ_z.

which $\theta_z > \theta_0$, then we can define an effective scattering absorption coefficient

$$\alpha_{sc}\left(\frac{\theta_0}{\theta_*}\right) = \int_{\theta_0}^{\pi} d\theta_z' \int_0^{\theta_z'} d\theta_y'\, \Sigma. \qquad (13.4.8)$$

This quantity is shown in Fig. 13.2 as a function of θ_0/θ_*. It is seen

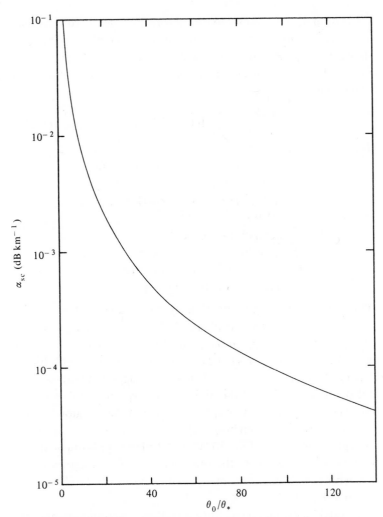

Fig. 13.2. Absorption coefficient, α_{sc} (13.4.8), in dB km^{-1} is shown as a function of the loss-cone angle, θ_0/θ_*.

that 'scattering loss' depends sensitively on the acceptance 'cone' chosen.

The term α_r in α_0 (13.3.5) also does not represent absorption, but just focussing effects associated with refraction. The term α_s, describing scattering by plankton, fish, etc. is expected to be very variable with location. This is not considered here.

The term α_d corresponds to true absorption due to hydrodynamic viscosity and ionic relaxation processes (related principally to $MgSO_4$ ions). The data relating to α_d above 1 kHz have been summarized in a semi-empirical formula by Schulken and Marsh (1962). This is

$$\alpha_d = \left(\frac{SAf_Tf^2}{f_T^2+f^2} + \frac{Bf^2}{f_T}\right)(1-6.54\times10^{-4}P_a)\,\text{m}^{-1} \qquad (13.4.9)$$

Here

$A \approx 2.34 \times 10^{-6}$, due to ionic relaxation,
$S =$ salinity (‰),
f_T is relaxation frequency (kHz)
$\quad \approx 21.9 \times 10^{[6-(1520/T+273)]}$,
$T =$ temperature in °C,
$f =$ acoustic frequency in kHz,
$B \approx 3.38 \times 10^{-6}$, due to viscosity,
$P_a =$ pressure in atmospheres.

$\qquad (13.4.10)$

Observations of attenuation (Sheely and Halley, 1957; Thorp, 1965; Kibblewhite and Denham, 1971; Mellen and Browning, 1975) over long propagation paths show considerable regional variability for frequencies below 1 kHz and are considerably higher than predicted from $\alpha_d + \alpha_{loss}$. It has been suggested (Kibblewhite and Bedford, 1975) that the observed absorption is influenced by differences in modes of propagation in different regions of the oceans. Fisher and Simmons (1977) have identified the extra absorption between about 50 Hz and 1 kHz with boric acid. For frequencies above 1 kHz the observed attenuation agrees rather well with (13.4.9). Values of α_d, expressed as dB km^{-1} loss, are shown in Fig. 13.3.

A model for estimating scattering loss from the sound channel has been given by Mellen et al. (1974). We now describe a slightly modified version of their analysis. Returning to (13.3.4) we neglect scattering in the horizontal direction, setting $\kappa_y = 0$. We also suppose that the acoustic energy flow is directed parallel to the x-axis and follows the ray path $z = z(x)$, $y = 0$. For $\theta_z > \theta_L$ it is supposed that energy is lost from the channel, so as a boundary condition on θ_z we set

$$I_\sigma(x, \theta_z) = 0 \quad \text{for } \theta_z \geq \theta_L. \qquad (13.4.11)$$

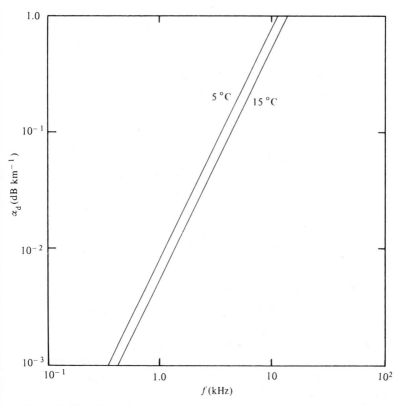

Fig. 13.3. The absorption coefficient, α_d (13.4.9), is shown as a function of the frequency, f, for seawater at temperatures of 5 °C and 15 °C.

Also, at $x = 0$, the chosen boundary condition is

$$I(0, \theta_z) = I_0 \quad \text{for } \theta_z < \theta_L$$
$$= 0 \quad \text{for } \theta_z > \theta_L. \tag{13.4.12}$$

To satisfy these boundary conditions a solution is assumed of the form

$$I(x, \theta_z) = \sum_{n=0}^{\infty} c_n \exp\left(-\int_0^x \beta_n \, dx'\right) \cos\left[\left(n + \frac{1}{2}\right)\pi \frac{\theta_z}{\theta_L}\right], \tag{13.4.13}$$

with

$$c_n = \frac{4}{\pi} \frac{(-1)^n}{(2n+1)} I_0. \tag{13.4.14}$$

To satisfy (13.3.4) we take

$$\beta_n = \pi^2 \left(n + \frac{1}{2}\right)^2 \frac{D_z}{\theta_L^2}. \tag{13.4.15}$$

At large distances the term corresponding to $n = 0$ will dominate the sum in (13.4.12). Using (13.4.4) we find that

$$\beta_0 = 6.4 \times 10^{-6} \frac{[n(z)/n(-B)]^5}{\theta_L^2} \frac{dB}{km}. \tag{13.4.16}$$

The tabulations of Munk (1974) suggest taking $\theta_L \approx 0.21$ for a ray on the axis. An estimate of the ray excursions off-axis gives

$$\alpha_{loss} = \frac{1}{x} \int_0^x \beta_0 \, dx' \approx 3.5 \times 10^{-4} \, dB \, km^{-1}. \tag{13.4.17}$$

Comparison of (13.4.17) with the values of α_d shown in Fig. 13.3 suggests that the scattering losses can be significant for frequencies less than ~ 50 Hz.

If the acceptance cone for detection is smaller than $\theta_L = 0.2$, as above, scattering losses can be very significant. This is seen from a comparison of Figs. 13.2 and 13.3.

The relations (13.4.5) provide simple estimates for the expected angular deflection of a ray due to internal-wave scattering. The mean increase in path length resulting from scattering may also be estimated from (13.4.5) as

$$\overline{\Delta s} \sim \frac{1}{2} \int_0^R \overline{\theta_z^2} \, dy \sim 3 \times 10^{-7} \, R^2, \tag{13.4.18}$$

expressed in km.

13.5. Scattering from the microstructure fluctuations

The observation of both a finestructure and a microstructure in temperature and salinity gradients was described in § 1.3. These will lead to a corresponding small-scale variation in sound speed. In contrast to the case of scattering by the gross structure, as discussed in § 13.2, the microstructure will give rise to large-angle scattering.

The observational data concerning this small-scale structure are not at present sufficiently complete to permit a detailed analysis of its contribution to the scattering of acoustic waves. It is likely that there are significant variations with depth and with location in the spectrum of these fluctuations. We shall attempt, therefore, to make only a crude estimate of the microstructure scattering. From the outset, we shall assume the acoustic wavelength to be long compared with the scale of the microstructure. By definition, the scale of the microstructure is less than one meter, so we are restricting ourselves to frequencies of less than 1.5 kHz.

With this condition we can rewrite (13.2.10) in the approximate form

$$\Sigma_m \approx \left(\frac{q^2}{2\pi}\right)^2 \int d^2x \langle \mu(0)\mu(\mathbf{x})\rangle_m, \qquad (13.5.1)$$

where the subscript 'm' implies that only the microstructure contribution is included. The integral here may be expressed in the form

$$\int d^2x \langle \mu(0)\mu(x)\rangle_m \equiv (\mu^2)_m l_H^2 l_V, \qquad (13.5.2)$$

where l_H and l_V are the respective horizontal and vertical scales of the microstructure fluctuations. Using the expressions (13.5.1) and (13.5.2), we can obtain the attenuation coefficient for microstructure scattering as

$$\alpha_m = \int d\Omega_{\hat{q}} \, \Sigma_m \approx 4\pi\Sigma_m$$
$$= \left(\frac{q^4}{\pi}\right) l_H^2 l_V (\mu^2)_m. \qquad (13.5.3)$$

From (1.1.5), we obtain

$$(\mu^2)_m = \langle (\alpha \, \partial_z T + \beta \, \partial_z S)^2 \rangle_m$$
$$\approx \alpha^2 \langle (\partial_z T)^2 \rangle_m. \qquad (13.5.4)$$

Three sets of observations are summarized in Table 1.4. The largest fluctuations are contained in the record MR7, which gives

$$(\mu^2)_m \approx 5 \times 10^{-7}. \qquad (13.5.5)$$

Precise values of l_H and l_V cannot be deduced from the data. We might estimate l_V to be somewhat less than 1 meter and l_H to be from 1 to 10 times l_V. For illustrative purposes we choose

$$l_H^2 l_V = 1 \text{ m}^3 \qquad (13.5.6)$$

and obtain from (13.5.3)

$$\alpha_m \approx 0.1 \, f^4 \text{ dB km}^{-1}, \qquad (13.5.7)$$

where f is the acoustic frequency in kHz.

A comparison of (13.5.7) with Fig. 13.3 suggests that α_m may be comparable to, or larger than, α_d when $f > 0.3$ and for fluctuations as large as those in record MR7. Had we chosen records MR6 or MSR7, on the other hand, the resulting coefficient α_m would have been two orders of magnitude smaller than (13.5.7) and of questionable importance for acoustic propagation studies.

It would appear from the discussion just given that microstructure may under some circumstances be of importance for acoustic propagation. A more precise evaluation of the importance of microstructure cannot at present be made, however.

PART V

EXPERIMENTAL OBSERVATIONS OF ACOUSTIC FLUCTUATIONS

In this part we describe several experiments, place them in the context of the general theory we have outlined in Parts III and IV, and compare their results quantitatively with theoretical expectations derived from internal-wave dominance.

Without attempting to be complete, we have chosen three experiments: one in the Western Atlantic (Eleuthera–Bermuda), one in the Eastern Atlantic (Azores), and one in the Northeastern Pacific (Cobb Seamount). Their characteristics are shown in Table V.1. The diffraction and strength parameters have been calculated for each experiment and are given in Fig. V.1. It is seen that the gamut from unsaturated, through partially saturated, to fully saturated is covered, although no experiment is comfortably far from boundaries.

The Eleuthera–Bermuda CW experiment (Chapter 14) is in the fully saturated region with many micromultipaths, in addition to a large number of deterministic rays. Therefore it is analyzed in terms of the multipath theory covered in Chapter 9.

Table V.1.

Characteristic	Bermuda	Cobb	Azores
Range	550 km, 1250 km	17.2 km	35 km
No. of deterministic paths	14, 34	1	1
Frequencies	406 Hz	4000, 8000 Hz	400–5000 Hz
Sampling interval	300 s	15.7 s	10 s
Λ	0.12, 0.24	$10^{-3}, 5 \times 10^{-4}$	0.06–0.005
Φ	13, 20	5, 10	10–100

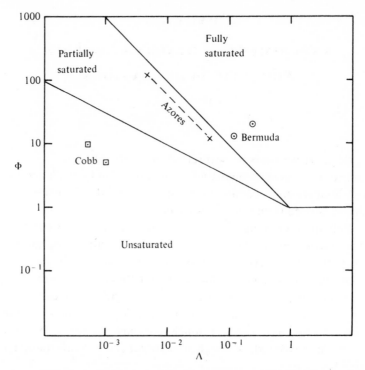

Fig. V.1. The set of analyzed experiments placed in the Λ–Φ diagram. The two points of the Bermuda experiment (Chapter 14) are at two different ranges; the two points of the Cobb experiment (Chapter 15) are at two different frequencies; the Azores experiment (Chapter 16) covers a band of frequencies.

The Cobb Seamount pulsed experiment (Chapter 15) is near the boundary of the unsaturated and partially saturated regions. The unsaturated theory is used for comparison.

The Azores experiment (Chapter 16) lies squarely in the most complicated part of the Λ–Φ diagram: the partially saturated region. It is thus more difficult to analyze, but correspondingly more rewarding. The analysis is based on the micromultipath theory of Chapters 11 and 12.

CHAPTER 14

ELEUTHERA–BERMUDA

A series of measurements of 406 Hz sound transmission between Eleuthera (Bahamas) and Bermuda was carried out over several years by a collaboration between the Universities of Miami and Michigan (the MIMI project; and referenced under MIMI). The long range of transmission (1250 km) put the experiment near the boundary of the fully saturated region, with many sporadic micropaths present. Phase fluctuations along any given path were large (one can estimate the rms phase fluctuation of a single path to be about 4 cycles). In addition, there were many deterministic paths from the source to receiver (Fig. 14.1). Therefore the conditions assumed in Chapter 9 are valid, and the only significant parameter to obtain from the data is ν, the rms phase rate averaged over the rays.

The path-integral treatment of the fully saturated region tells us that the phase rate for each ray can be calculated as $\dot{\Phi}$ even when sporadic micropath is present (Chapter 12). Therefore results of Chapter 7 for $\dot{\Phi}$ from internal-wave dominance are applicable and are compared with experiments below (§ 14.2). This comparison was previously reported by Munk and Zachariasen (1976). The treatment of phase and intensity statistics (§ 14.3) was previously discussed by Dyson, Munk, and Zetler (1976).

14.1. Treatment of data

The observational material, generously made available to us by John G. Clark, consists of intensity $I(t)$ (dB, arbitrary reference) and phase $\phi(t)$ (in cycles), as represented by the top two curves of Fig. 14.2; a selected portion is shown point-by-point in Fig. 14.3. A similar set of data from an intermediate receiver has also been used (Mid-station, nominal range 550 km). The observed acoustic

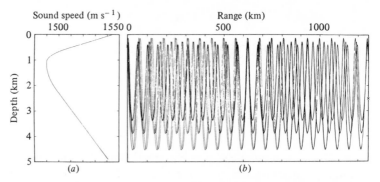

Fig. 14.1. (*a*) The mean sound profile for MIMI transmission, and (*b*) the associated ray diagram. For clarity we show only every other ray doublet, altogether 6 rays out of a total of 13. We are indebted to John Clark of the Institute for Acoustic Research for this figure.

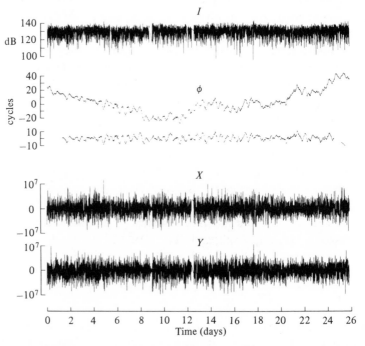

Fig. 14.2. Plots of measured intensity I, phase ϕ and highpassed phase, and of the components X and Y (in arbitrary pressure units) of acoustic pressure at Bermuda, 22 September to 17 October 1973. (From Dyson, Munk and Zetler, 1976.)

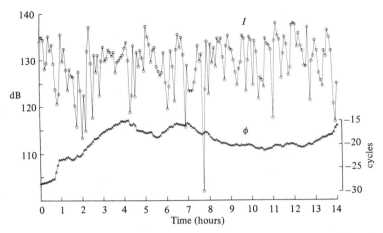

Fig. 14.3. A 14-hour sample of intensity I and phase ϕ drawn on an enlarged scale. (From Dyson, Munk and Zetler, 1976.)

pressure fluctuations (frequency $\sigma = 406$ Hz) relative to some (unspecified) scale p_0 can be written

$$p(t)/p_0 = x(t) \cos \sigma t + y(t) \sin \sigma t$$

where x, y are slowly varying (compared to σ) amplitudes. The original measurements consist of the five-minute averages of amplitudes

$$X(t) = \frac{1}{\delta t} \int_0^{\delta t} x(t) \, dt = A(t) \cos \phi(t), \quad Y(t) = A(t) \sin \phi(t).$$

(14.1.1)

Intensities are defined as in (8.1.5) to (8.1.7) and $\phi(t)$ is the phase.

Only fractional cycles are measured, and there is an ambiguity concerning the integer number of cycles. Normally this can be resolved by the continuity of the time series. Phase difference over the sampling interval $\delta t = 5$ m has an rms value of $\delta\phi = 0.24$ cycle. The probability for $\delta\phi = \phi_i - \phi_{i-1}$ to exceed $\frac{1}{2}$ cycle is 4 per cent (8 per cent were observed). A restriction to $|\delta\phi| \leq \frac{1}{2}$ cycle (which can be attained by adding and subtracting integer cycles) is not realistic. We have edited the observations to remove phase 'kinks', replacing the reported value ϕ by $\phi \pm 1, \phi \pm 2, \ldots$ cycles when required to make the adjusted phase difference $\delta\phi_n = \phi_{n+1} - \phi_n$ subject to the

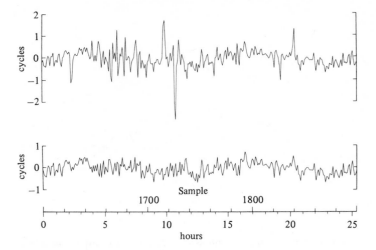

Fig. 14.4. Bermuda phase differences, $\delta\phi$ before adjustment (a) and after adjustment (b). (From Dyson, Munk and Zetler, 1976.)

restriction

$$|\delta\phi_n - \tfrac{1}{2}(\delta\phi_{n+1} + \delta\phi_{n-1})| \leq \tfrac{1}{2} \text{ cycles.} \qquad (14.1.2)$$

This is essentially placing an upper limit on second differences in phase; 5 per cent of the Bermuda observations and 1.5 per cent for the Mid-station were adjusted accordingly. Fig. 14.4 shows a sample of $\delta\phi$ before and after adjustment, and Fig. 14.5 the reconstituted $\phi = \sum \delta\phi$. Mid-station phases are not severely altered by phase adjustments. At Bermuda the low (week-to-week) frequencies bear no resemblance to the Mid-station trend, and are severely altered by the phase adjustment; however, the highpassed records (tidal frequencies and higher) are not severely altered. We conclude that recorded Mid-station phases and high-frequency Bermuda phases are significant, but that sampling was not adequate to obtain low-frequency trends at Bermuda. Adjusted mean-square differences and phase differences are given in Table 14.1.

Multipath intensities are characterized by occasional deep fades (Figs. 14.2 and 14.3), and it will be useful to censor these selectively. To be specific, we replace the recorded values of I by $I_0 - F$ whenever $I < I_0 - F$, but otherwise leave I unchanged; censored X and Y are subsequently computed according to (14.1.1). Accord-

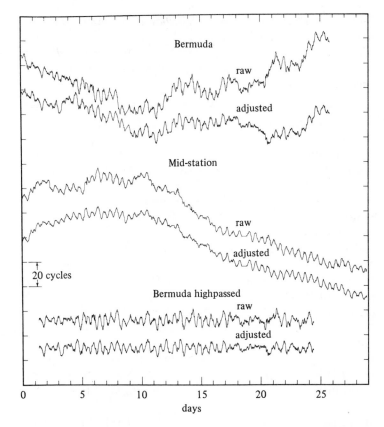

Fig. 14.5. Bermuda and Mid-station phases before and after adjustment. (From Dyson, Munk and Zetler, 1976.)

ingly the three columns in Table 14.1 refer to the removal of fades exceeding $F = \infty$, 20, 10 dB (the first column then refers to the uncensored record). A typical signal-to-noise ratio is 27 dB (Ted Birdsall, personal communication), a removal of deeper fades (say $F = 30$ dB) would be associated with noise statistics rather than multipath statistics.

14.2. Cartesian spectra

Munk and Zachariasen (1976) have calculated the Cartesian spectra, that is, the Fourier transforms of $\langle X^2 \rangle$ and $\langle Y^2 \rangle$, for comparison

Table 14.1. *Phase and intensity statistics at* 406 Hz *for data interval* $\delta t = 5.0394$ *minutes.* X, Y, $R = (X^2 + Y^2)^{\frac{1}{2}}$ *in arbitrary pressure units. The columns marked* ∞, 20, 10 *correspond to the suppression of fades exceeding* ∞ (*no change*), 20, 10 dB, *respectively.*

Range from Eleuthera (km)	Mid-station			Bermuda		
Record length (days)	550 (nominal)			1250		
	8255 terms = 29 days			7366 terms = 26 days		
$\langle(\delta\phi)^2\rangle$ (cycles^2)	0.029			0.060		
F (dB)	∞	20	10	∞	20	10
Number of terms replaced	0	114	1076	0	79	746
$\langle X^2\rangle \times 10^{-13}$	8.73	8.73	8.79	0.668	0.667	0.671
$\langle Y^2\rangle \times 10^{-13}$	8.70	8.70	8.76	0.655	0.655	0.659
$\langle(\delta X)^2\rangle \times 10^{-13}$	6.08	6.08	6.14	0.797	0.797	0.806
$\langle(\delta Y)^2\rangle \times 10^{-13}$	5.92	5.92	5.98	0.790	0.791	0.799
$\langle R^2\rangle \times 10^{-13}$	17.43	17.43	17.55	1.323	1.322	1.330
I_0 (dB)	142.41			131.22		
$\langle I\rangle$ (dB)	139.22	139.27	139.82	128.61	128.65	129.06
$\langle I^2\rangle - \langle I\rangle^2$ (dB^2)	38.79	36.20	23.01	31.84	30.20	21.13
$\langle(\delta I)^2\rangle$ (dB^2)	24.10	21.77	13.09	48.17	45.31	30.34
$\langle\lvert \delta I \cdot \delta\phi\rvert\rangle$ (dB · cycles)	0.55	0.52	0.38	1.15	1.13	0.92

with theory. Fig. 14.6 shows the spectra with plots of (9.2.3) for the indicated values of ν^{-1}. One sees that the spectra have the specific shape expected from the (rather drastic) assumptions of Chapter 9. Experimental values of ν can be deduced from Fig. 14.5, or from the formula

$$\nu^2 = \frac{\langle\dot{X}^2\rangle + \langle\dot{Y}^2\rangle}{\langle X^2\rangle + \langle Y^2\rangle}. \qquad (14.2.1)$$

The values so inferred for Mid-Station and Bermuda are given in Table 14.2 as ν (experimental).

A theoretical value for ν^2 may be calculated assuming internal-wave dominance by use of the results of Chapter 7. We recall from

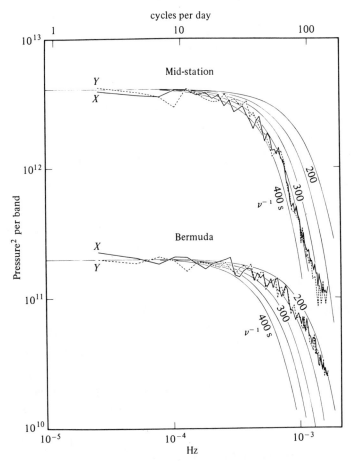

Fig. 14.6. Spectra of the Cartesian pressure components X, Y (in arbitrary units) per bandwidth 2.58×10^{-5} Hz. The computed curves are drawn for indicated values of ν^{-1} in seconds. (From Dyson, Munk and Zetler, 1976.)

(9.2.3) that $\nu = \dot{\Phi}$ and that $\dot{\Phi}$ is given in Chapter 7 in terms of internal-wave parameters (§ 7.5).

The experiment involves a large number of deterministic rays. Therefore we present results for two extremes (surface-limited ray and axial ray), and then give a weighted average over an appropriate ray mix. For the *surface-limited ray* (7.5.17) and (7.4.8) yield

$$\nu^2 = \dot{\Phi}_a^2 = \omega_i n_0 q^2 BR \langle \mu_0^2 \rangle \langle j^{-1} \rangle (8/\pi^2)[(2\varepsilon r/3\pi B)^{\frac{1}{2}} \ln (n_0/\omega_i)]$$

$$(14.2.2)$$

Table 14.2. *Comparison of ν values from experiment with the theory of internal-wave dominance.*

	Mid-station	Bermuda
Range (km)	550 (nominal)	1250
Paths i	14	34
Double-loops		
Surface-limited ray (SLR)	10	22
Near-axial ray	13	30
$ν$ (Experimental)	$2.8 \times 10^{-3} \, \text{s}^{-1}$	$4.0 \times 10^{-3} \, \text{s}^{-1}$
$ν$ (Internal-wave dominance)		
Apex approximation, SLR	$4.6 \times 10^{-3} \, \text{s}^{-1}$	$6.8 \times 10^{-3} \, \text{s}^{-1}$
Axial ray	$2.9 \times 10^{-3} \, \text{s}^{-1}$	$4.4 \times 10^{-3} \, \text{s}^{-1}$
Weighted average	$3.5 \times 10^{-3} \, \text{s}^{-1}$	$5.2 \times 10^{-3} \, \text{s}^{-1}$

while for the *flat ray* (7.5.16) and (7.4.5) give

$$\nu^2 = \dot{\Phi}_f^2 = \omega_i n_0 q^2 BR \langle \mu_1^2 \rangle \langle j^{-1} \rangle (8/\pi^2) \ln (n_1/\omega_i). \quad (14.2.3)$$

Parameters from the internal-wave model of Chapter 3 lead to the values in Table 14.2.

A more precise calculation may be done (Munk and Zachariasen, 1976) allowing for the proper ray mix. The result may be expressed in the following form:

$$\nu^2 = \omega_i n_0 q^2 BR_D \langle \mu_0^2 \rangle \langle j^{-1} \rangle \sum_i \Delta_i (K^+ F_2^+ + K^- F_2^-)_i \quad (14.2.4)$$

where $K^{\pm}$ is the number of upward or downward loops and Δ_i is a weighting factor for the ith ray. Let $\bar{\theta}_i$ be the axis-crossing angle of the ith ray. Then

$$\Delta_i = (\bar{\theta}_i - \bar{\theta}_{i-1})/\sum_i (\bar{\theta}_i - \bar{\theta}_{i-1}). \quad (14.2.5)$$

The various terms for each ray are given in Table 14.3. The summand (last column) is fairly uniformly distributed among all contributing rays. Computed and acoustically measured values are within 24%: rather too close considering the uncertainty in the ocean model. We conclude that internal waves consistent with

Table 14.3. *Calculations of* v^2 *for Bermuda.* $\bar{\theta}_i$ *are the inclinations at the axial source of all possible rays to an axial receiver at 1250 km range, consisting of* K^+ *upper loops of range* R^+, *and* K^- *lower loops of range* R^- *(see Fig. 14.1).* $F_2^\pm(0)$ *are the dimensionless contributions per ray loop to* v^2, *leading to the dimensionless weighted sum* $\Delta_i(K^+F_2^+ + K^-F_2^-)_i$.

$\bar{\theta}_i$	K^+	R^+	K^-	R^-	$\Delta\theta$	F_2^+	F_2^-	Summand
12.7	22	12.8	22	44.0	0.60	0.703	0.011	0.362
12.3	23	12.9	22	43.3	0.60	0.680	0.012	0.367
11.5	23	13.2	23	41.2	0.60	0.630	0.014	0.342
11.2	24	13.4	23	40.4	0.55	0.614	0.015	0.319
10.3	24	13.7	24	38.4	0.55	0.560	0.018	0.293
10.0	25	13.9	24	37.6	0.55	0.545	0.019	0.298
9.2	25	14.2	25	35.8	0.55	0.500	0.021	0.276
9.0	26	14.4	25	35.0	0.55	0.490	0.022	0.281
8.1	26	14.8	26	33.3	0.60	0.448	0.028	0.286
7.8	27	15.1	26	32.4	0.60	0.435	0.029	0.288
6.9	27	15.5	27	30.8	0.60	0.395	0.035	0.268
6.5	28	15.9	27	29.8	0.65	0.375	0.038	0.288
5.7	28	16.4	28	28.2	0.70	0.343	0.045	0.292
5.1	29	16.7	28	27.4	0.85	0.315	0.050	0.344
4.0	29	17.5	29	25.6	1.05	0.274	0.062	0.394
3.0	30	18.2	29	24.3	1.50	0.235	0.073	0.529
1.0	30	19.8	30	21.9	2.00	0.165	0.107	0.628
−1.0	30	19.8	30	21.9	1.80	0.165	0.107	0.628
−2.6	29	18.5	30	23.8	1.50	0.219	0.078	0.501
−4.0	29	17.5	29	25.6	1.10	0.274	0.062	0.394
−4.8	28	16.8	29	26.9	0.85	0.300	0.054	0.326
−5.7	28	16.4	28	28.2	0.70	0.343	0.045	0.292
−6.2	27	16.0	28	29.2	0.65	0.365	0.040	0.274
−6.9	27	15.5	27	30.8	0.60	0.395	0.035	0.268
−7.4	26	15.3	27	31.6	0.60	0.418	0.031	0.270
−8.1	26	14.8	26	33.3	0.55	0.448	0.028	0.286
−8.5	25	14.7	26	33.9	0.55	0.469	0.025	0.262
−9.2	25	14.2	25	35.8	0.55	0.500	0.021	0.276
−9.6	24	14.0	25	36.6	0.55	0.512	0.020	0.271
−10.3	24	13.7	24	38.4	0.55	0.560	0.018	0.293
−10.6	23	13.6	24	39.0	0.55	0.585	0.016	0.293
−11.5	23	13.2	23	41.2	0.60	0.630	0.014	0.342
−11.8	22	13.1	23	41.8	0.60	0.652	0.013	0.338
−12.7	22	12.8	22	44.0	0.60	0.703	0.011	0.362

oceanographic observations can account for the measured acoustic fluctuations.

14.3. Phase and intensity statistics

The observed data (for example, Figs. 14.2 and 14.3) are customarily plotted in terms of intensity and phase. We are therefore interested in calculating the statistical behavior of ι and ϕ predicted by the theory of Chapter 9.

We begin with average values. I_{rms} is predicted to be 5.57 dB (8.1.9), as compared to the measured values of 6.2 and 5.6 dB for Mid-station and Bermuda, respectively. From (9.3.1), the correlation C is predicted to be 0.63, as compared to 0.66 and 0.68 for Mid-station and Bermuda. The properties of fadeouts predicted by (9.3.2) to (9.3.5) cannot be compared with experiment because of the five-minute averaging present in the data. All one can say is that the results do not contradict the computation, but for adequate studies one will need to sample at least once per minute.

Fig. 14.7 shows the comparison between the computed spectra and the observed spectra. The overall agreement is not good, though the predicted ω^{-1} and ω^{-3} roll-offs for rate-of-phase and intensity spectra are borne out at Mid-station. The high-frequency Bermuda intensities are aliased from undersampling. There are discrepancies at the lowest frequencies, some of which can be attributed to tides.

Mean-square variations computed from internal waves are:

$$\langle \dot{\phi}^2 \rangle = \int_0^{\omega'} F_{\dot{\phi}}(\omega)\, d\omega = \tfrac{1}{2}\nu^2 \sinh^{-1}(2\omega'/\nu)$$

$$\langle \iota^2 \rangle = \int_0^{\omega'} F_\iota(\omega)\, d\omega = 2\nu^2[\sinh^{-1}\alpha - \alpha(1+\alpha^2)^{-\frac{1}{2}}],$$

$$\alpha = (\pi/2\sqrt{3})\,\omega'/\nu$$

(14.3.1)

and these become logarithmically infinite as $\omega' \to \infty$. The upper limit is set by the integration time δt, and crudely $\omega' = 2\pi/\delta t$. Results are given in Table 14.4.

14.4. Conclusions

The measured acoustical fluctuations for project MIMI are close to the expected values computed for an internal-wave spectrum based

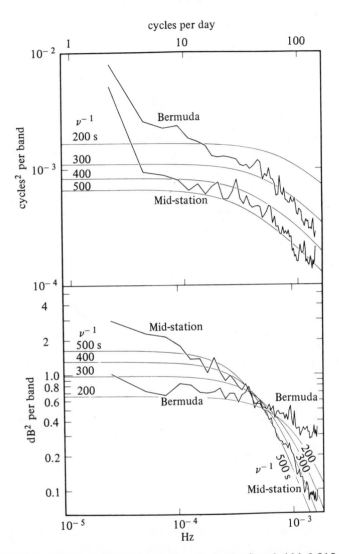

Fig. 14.7. Spectra of phase *difference* and intensity (bandwidth 0.915 cpd). The computed curves are drawn for indicated values of ν^{-1} in seconds. The area under the intensity spectra (e.g., the mean-square fluctuations) is independent of ν. (From Dyson, Munk and Zetler, 1976.)

Table 14.4. *Rms phases and intensities* $(\omega' = 0.0208\,s^{-1})$.

	Mid-station $(\nu^{-1} = 357\,s)$		Bermuda $(\nu^{-1} = 250\,s)$	
	rms $\delta\Phi$	rms δI	rms $\delta\Phi$	rms δI
Theory of internal-wave dominance	0.19 cycles	7.0 dB	0.26 cycles	8.9 dB
Observed	0.17	4.9	0.25	6.9

entirely on oceanographic observations. The agreement could be improved by reasonable adjustments of the internal-wave parameters, but this is not the object. The MIMI situation is consistent with the existence of many deterministic multipaths associated with the sound channel; the result is not sensitive to additional sporadic multipaths associated with the internal-wave perturbations.

The tidal contribution to the acoustic fluctuations has been emphasized in the literature, perhaps because of a superficial resemblance of the phase fluctuation $\phi(t)$ to tidal records (Fig. 14.5). Our conclusion is that tides play a significant but not dominant role. Detailed analysis of the tidal effects is contained in Dyson, Munk and Zetler (1976).

Overall, agreement of theory and experiment is good for Cartesian averages and spectra, and for intensity-phase averages; but intensity-phase spectra have discrepancies which may be due to fluctuations that are coherent between a large number of separate rays.

14.5. The Williams and Battestin resolved experiment

Williams and Battestin (1976) have measured phases and phase spectra in the Bermuda area, using selected paths resolved from others in the multipath set. The resolution was accomplished by means of a long bottom-moored vertical array with sufficient vertical directivity to separate most arrivals, and this was supplemented by an arrival time resolution of 10 ms. For typical ranges of 500 km, a typical rate of phase at 400 Hz is $3 \times 10^{-3} \, s^{-1}$. This is about the same as the MIMI results (Table 14.2).

Offhand, the result is surprising; we would have expected the singlepath $\nu^2 = \langle \dot{\phi}_i^2 \rangle$ to be much lower than the multipath $\langle \dot{\phi}^2 \rangle$. In fact from (14.3.1) the ratio $\langle \dot{\phi}^2 \rangle / \nu^2 = \frac{1}{2} \sinh^{-1}(2\omega'/\nu)$ becomes logarithmically infinite as $\omega' \to \infty$. The upper limit is set by the integration time δt, and crudely $\omega' = 2\pi/\delta t$. But for conditions under which this experiment was conducted (fully saturated) the micromultipaths are typically separated by more than a vertical coherence distance, and so we are back to multipath statistics, and phase rates of the resolved macropaths should be the same as those of the unresolved macropaths, as observed.

This satisfactory conclusion is somewhat tempered by the fact that the acoustic source was freely drifting, so that the measured phases had first to be corrected for variable range, using polynomial fitting to LORAN C positions.

CHAPTER 15

COBB SEAMOUNT

Ewart (1976) has measured amplitude and phase (transit time) fluctuations between a fixed transmitter and receiver on Cobb Seamount (46°46′ N, 130°47′ W). The sound axis is shallow, 400 m, as is characteristic of high latitude. Setting $z_1 = -0.4$ km in (1.6.6), we construct a ray path (Fig. 15.1) through source and receiver (both at 1000 m depth) separated by 17.2 km, with a lower turning point at a depth of 1350 m, in agreement with ray tracing based on locally measured sound profiles (Fig. 4 in Ewart, 1976). Further, the measured $n(z)$ is very close to our exponential model (1.6.3). Ewart obtained 144.5 hours of records (with minor gaps) based on 8-cycle pulses at 4166 Hz and 16-cycle pulses at 8333 Hz transmitted alternately every 15.7 s.

15.1. Phase and intensity variances

The measured rms variations are given in Table 15.1. Ewart remarks on the strong tidal contribution to the travel time spectra, and on the important effect on intensity by sporadic multipaths associated with sound velocity finestructure. We note that the results are similar at the two frequencies (as expected); the rms phase at 4 KHz is 3.84×10^{-4} s $\times$ 4166 Hz = 1.60 cycles.

The computed values of Λ and Φ (Table V.1) place this experiment in the geometric-optics region, close to the partially saturated region (Fig. V.1). The numerical value of Φ follows from the analyses of Chapter 10;

$$\Phi^2 = q^2 B R_D \langle \mu_0^2 \rangle \langle j^{-1} \rangle F_1(0) = 22.8, \qquad (15.1.1)$$

where we have used $z_1 = -0.4$ km, $B = 1$ km, so that rms $\mu_0 = 4.9 \times 10^{-4} \exp 2(-z_1/B - 1) = 1.5 \times 10^{-4\dagger}$ (1.6.8); further $\langle j^{-1} \rangle =$

†Munk and Zachariasen (1976) incorrectly use $z_1 = -B$.

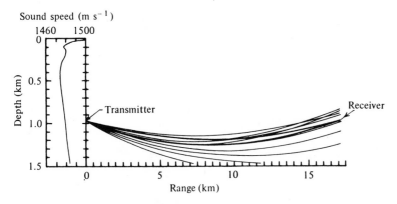

Fig. 15.1. Sound-speed profile and ray geometry in the Cobb Seamount experiment. (From Ewart, 1976.)

Table 15.1. *The measured rms values of travel time and intensity.*

Frequency (nominal)	4 kHz	8 kHz
Travel time	0.384 ms	0.374 ms
Intensity	2.3 dB	2.4 dB

0.435, $q = 1.745 \times 10^4$ rad km^{-1} (for 4166 Hz), $R = 20.8$ km, $n_0 = 5.2 \times 10^{-3}$ s^{-1}, $\omega_i = 1.06 \times 10^{-4}$ s^{-1} (46.75° latitude), and $F_1(0) = 0.38$ (from a numerical integration). Thus $\phi_{rms} = 0.76$ cycles, compared to 1.60 measured.

Similarly the intensities are found from (10.4.26) using $A_{zz}^{-1} = x(R-x)/R$ which is appropriate for a single lower loop. The result is $\langle \iota^2 \rangle - \langle \iota \rangle^2 = 0.022$.

(A more accurate form for A_{zz}^{-1} will increase the calculated value slightly.) The observed value of 2.3 dB for log-intensity fluctuations corresponds to a variance of log-amplitude of

$$\langle \iota^2 \rangle - \langle \iota \rangle^2 = (2.3 \text{ dB})^2 (\log 10)^2 / 400 = 0.070$$

at 4 kHz and a very similar number at 8 kHz. The comparisons

Table 15.2. *Observed and computed values of phase and amplitude variations at* 4 kHz.

Quantity	Observed	Computed
Variance of phase	100	23
Variance of log-amplitude	0.070	0.022
Phase spectral form	ω^{-3}	ω^{-3}
Log-amplitude spectral form	$\sim\omega^{-1}$	ω^{-3}

between observed and computed variances are summarized in Table 15.2. Thus the computed variances are low by a factor of three or four.

15.2. Spectra

Measured and computed spectra are shown in Fig. 15.2. Munk and Zachariasen (1976) have calculated theoretical curves for these spectra assuming internal-wave dominance and making use of the Rytov approximation. Desaubies (1976) has treated the phase spectrum only, using the geometrical-optics approximation. The computed curves in Fig. 15.2 follow the Rytov method of Chapter 10; these curves are lower than those of Fig. 7 in Munk and Zachariasen (1976) by a factor $(3.3)^2$ in accordance with their footnote 13. The computed phase spectrum is somewhat low, as expected from the rms values, but in the principal band between inertial and buoyancy frequencies the computed ω^{-3} slope is reasonably consistent with the observed spectral slope. The observed phase spectrum continues smoothly beyond the computed n cutoff. Computed intensities completely fail to account for the observed high-frequency fluctuations.

Micromultipath cannot provide the measured intensity fluctuations, for even the observed result (log-amplitude fluctuations of 0.07) implies that the Rytov approximation should be valid. The conclusion is that internal waves cannot account for the measured intensity fluctuations. As discussed in Chapters 8 and 12, a very small horizontal current convecting the internal waves could

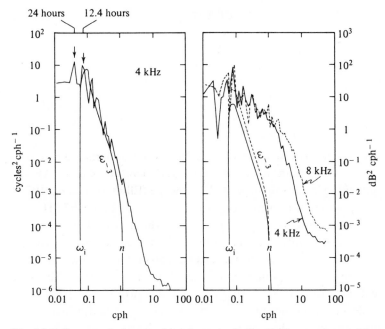

Fig. 15.2. Spectra of phase and log intensity in the Cobb experiment. The theoretical curves are computed from the internal-wave model using the Rytov method (Chapter 10). (From Munk and Zachariasen, 1976.)

substantially raise the high-frequency contribution to intensity fluctuations.

More detailed comparisons of theory with the Cobb Seamount experiment should await a careful removal of tidal effects from the spectra in Fig. 15.2, a subtraction that has not yet been done.

CHAPTER 16

AZORES

In March 1975, a series of acoustic and oceanographic measurements were made at the *Azores Fixed Acoustic Range* (AFAR) under the leadership of A. W. Ellinthorpe (Ellinthorpe *et al.*, 1977). Frequencies of transmission varied between 412 and 4672 Hz, and measurements were taken over a 120-hour period. We concern ourselves mainly with data obtained over a 35 km path from a source at 550 m depth to a receiver at 750 m. The local sound-speed profile leads to a single, fully refracted path (Fig. 16.1) which we calculate to be in the partially saturated region. Other paths which reflect from surface or bottom have been eliminated from the data by using pulses short enough to resolve them.

Direct measurements of the local temperature, salinity and sound speed were taken simultaneously with the acoustic measurements (§ 16.1). Gaussian-envelope acoustic pulses were transmitted at repetition intervals of 10.2 s. Pulses centered at several frequencies with widths (between −6 dB points) of 130 ms were used to simulate CW transmission (§ 16.2). Pulses centered at 3200 Hz with 1.28 ms width were used to obtain information about wander and spread in transmission time (§ 16.3).

16.1. Environmental data

Profiles of temperature, salinity, and sound speed were measured from R/V *Bannock* (Italy), B. O. *D'Entrecasteaux* (France) and R/V *Planet* (Federal Republic of Germany) at 36° 52.5′ N, 23° 35.8′ W (Fig. 16.2). The maximum in sound speed between 800 and 1000 m is the result of the Mediterranean intrusion, and is incompatible with a canonical sound-speed profile. The AFAR group has fitted the sound-speed profile between 240 m and 840 m

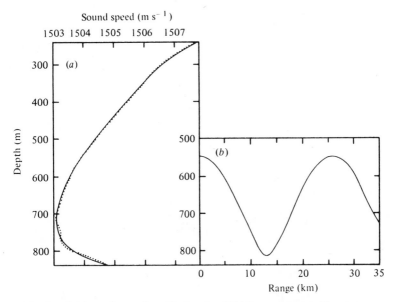

Fig. 16.1. (a) Sound-speed profile for the AFAR experiment. The dots are data averaged from 57 profiles. The smooth curve is a sixth-order polynomial fit. (b) Ray geometry over a 35 km path.

to a sixth order polynomial (Fig. 16.1) which we use in our analysis. The buoyancy frequency has also been calculated (Fig. 16.3).

Fluctuations of sound speed were measured by taking the variance of many profiles (Fig. 16.4). If the fluctuations are dominated by internal waves, then the fluctuations at different depths z and z' should be related by

$$\frac{\langle \mu^2(z) \rangle}{\langle \mu^2(z') \rangle} = \frac{n(z')}{n(z)} \left(\frac{\partial_z C_P(z)|_z}{\partial_z C_P(z)|_{z'}} \right)^2 \qquad (16.1.1)$$

where the first ratio (of buoyancy frequency) is an approximation to the relative *displacement* variance of internal waves, and the second ratio (of potential sound-speed gradient squared) represents the effect of a given displacement on sound-speed fluctuation (see (7.4.4)). Since $\partial_z C_P = \partial_z C + C_1 \gamma_A$ we may use our sixth-order polynomial fit to $C(z)$, and a smoothed version of the $n(z)$ shown in Fig. 16.3 to calculate $\langle \mu^2(z) \rangle$ within an arbitrary constant. The dashed line of Fig. 16.4, which is normalized to $\langle \mu^2 \rangle = 10^{-7}$ at 600 m depth, shows that this internal-wave prediction of the depth

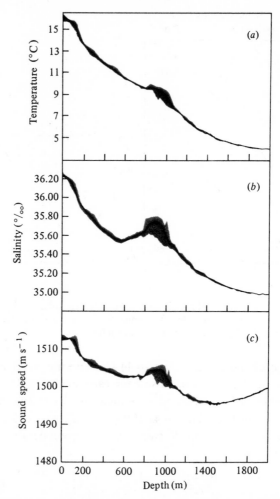

Fig. 16.2. Profiles of (*a*) temperature, (*b*) salinity, and (*c*) sound speed for the AFAR experiment. The shaded areas represent the spread in measurements.

dependence of $\langle \mu^2 \rangle$ is well verified in the region between 240 m and 740 m. Qualitatively it is seen that at the depth where $C^{-1}\partial_z C \approx -\gamma_A$; that is, where the sound-speed gradient is equal to the adiabatic gradient, $\langle \mu^2 \rangle$ is predicted to be zero, since an internal-wave displacement at this point causes no sound-speed fluctuation. This is calculated to occur at a depth of approximately 790 m.

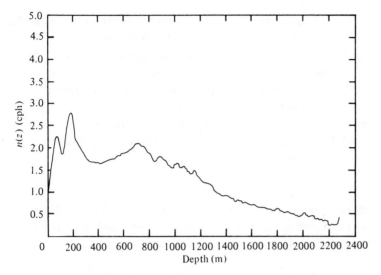

Fig. 16.3. Profile of buoyancy frequency in the AFAR experiment, determined from temperature and salinity measurements.

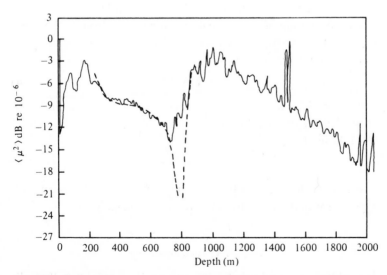

Fig. 16.4. Sound-speed fluctuation variance as a function of depth. The dashed line is a prediction of the internal-wave model, normalized to fit the data at 600 m depth.

Because of the mixing that must be occurring near 800 m due to the intrusion of Mediterranean waters, and because of the neglect of internal-wave water velocity effects compared with vertical displacement effects, the internal-wave predictions near 790 m are unreliable. However, the unperturbed ray spends relatively little time below a depth of 740 m, never goes below 810 m, and $\langle \mu^2 \rangle$ between 740 m and 810 m is relatively small, so that we feel justified in using the internal-wave predictions for AFAR as a first approximation.

Λ–Φ space

Only environmental data are needed to calculate Λ and Φ. From our expression for $\langle \mu^2 \rangle$ and the internal-wave expression for $L_P(\theta, z)$ we may calculate Φ^2 according to (7.2.4) and (7.4.1);

$$\Phi^2 = 4\pi^{-2}\langle j^{-1} \rangle (n_0/\omega_i) q^2 B \int_0^R \langle \mu^2 \rangle f_1(n\theta/\omega_i)\, ds. \quad (16.1.2)$$

A numerical evaluation of the integral along the ray results in

$$\Phi = \sigma/40 \text{ Hz} \quad (16.1.3)$$

where σ is the acoustic frequency, expressed in Hz.

The calculation of Λ may be done numerically from (7.2.5), or a reasonable approximation may be obtained from assuming no sound-channel effect, using

$$\Lambda = R/6qL_V^2 \quad (16.1.4)$$

with $L_V = 240$ m from (7.4.2) or from measurements by Ellinthorpe et al. (1977) for depths smaller than 740 m. An evaluation of (16.1.4) yields

$$\Lambda = 25 \text{ Hz}/\sigma. \quad (16.1.5)$$

We have, therefore, $\Lambda\Phi = 0.63$ and $\Lambda\Phi^2 = \sigma/64$ Hz placing the experiment in the partially saturated region (Fig. V.1).

16.2. CW measurements

The long-pulse (quasi-CW) measurements have been used to measure distributions of intensity (one-point functions) and correlations in time (two-point functions).

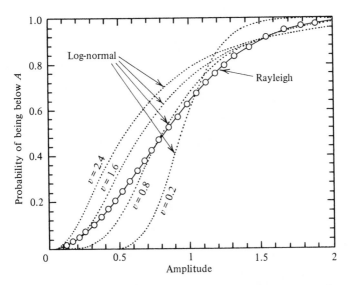

Fig. 16.5. Amplitude distribution in the 35 km AFAR experiment, compared with a Rayleigh distribution, and several log-normal distributions. All curves have unity mean intensity. The quantity v is the variance of log-intensity. The Rayleigh distribution, which has $v = 1.64$, is expected in this case. The open circles are experimental data.

One-point functions

The expectation that the amplitude is approximately Rayleigh-distributed in the partially saturated region is verified in Fig. 16.5. For comparison, the results from a 3 km-path experiment (Ellinthorpe *et al.*, 1977) in the unsaturated region follow the expected log-normal distribution in Fig. 16.6. From the intensity distributions the rms values of log-intensity may be calculated, and are found to increase with $\Lambda\Phi^2$ up to an asymptotic value of $\pi^2/6$ in the saturated region (Fig. 16.7).

The expected distribution is not expected to be exactly Rayleigh, even in the saturated region. The first correction term to the moments is given by (Chapter 12):

$$\langle I^n \rangle = n!\langle I \rangle^n (1 + \tfrac{1}{2}n(n-1)\gamma) \qquad (16.2.1)$$

where $\gamma = \tfrac{3}{2}\ln\Phi$ for a straight-line ray. Table 16.1 illustrates that the moments at 4672 Hz are better represented by (16.2.1) than by an exact Rayleigh distribution (for which $\gamma = 0$). The value of γ is

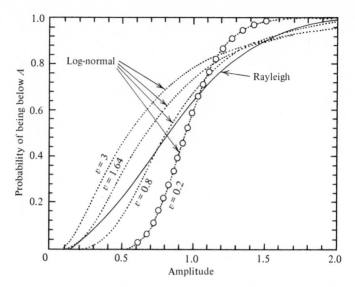

Fig. 16.6. Amplitude distribution in a 3 km AFAR transmission. (See Fig. 16.5 for the definition of v.) The value of $\Lambda\Phi^2$ is 0.2 for these data, so that a log-normal distribution with $v = 0.2$ is expected. The open circles are experimental data.

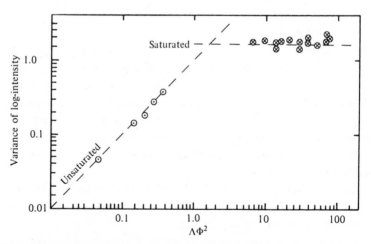

Fig. 16.7. Log-intensity variance as a function of $\Lambda\Phi^2$. The dots and crosses came from 3 km and 35 km experiments respectively with several different frequencies from 400 to 5000 Hz. The 45° dashed line is the expected result in the unsaturated region; the saturation value of $\pi^2/6$ is expected to be approached near $\Lambda\Phi^2 \approx 1$.

Table 16.1. *Intensity moments at*
4672 Hz. Calculated values are from
(16.2.1) with the value $\gamma = 0.2$ *chosen to*
fit the observed difference for n = 2.

n	Observed $\langle I^n \rangle$	Rayleigh prediction $n!\langle I \rangle^n$	$(\gamma = 0.2)$ calculated
1	1.0	1	1.0
2	2.4	2	2.4
3	9.4	6	9.6
4	52.7	24	52.8

difficult to calculate in the Azores case since the ray is not a straight line (Fig. 16.1). The experimental value of 0.2 is not far from the straight line approximation of 0.3. We cannot make an adequate comparison at lower frequencies because the total experiment time of 120 hours is insufficient to obtain the higher moments. (The higher moments are increasingly dominated by excursions to high intensity, which become increasingly infrequent at low frequencies.)

Two-point functions

Correlations between the wavefunctions at two points separated by time Δt are sensitive to the important environmental parameters. In all regions (8.4.5) is expected to be valid, and Fig. 16.8 verifies that for small time separations $D(t) \approx \dot{\Phi}^2 (\Delta t)^2$. The values of $\dot{\Phi}$ obtained at different frequencies may be used to estimate the environmental parameter $\Phi/\dot{\Phi}$ (Table 16.2), to be compared with the expectation from internal waves (7.5.16) of 1.4 hours.

The above result from the autocorrelation of $\psi(t)$ should be valid whatever the regime of the experiment. To distinguish between the partially and fully saturated region, as discussed in § 8.4, we consider the correlations listed in Table 16.3. The experimental results are shown in Fig. 16.9 where we have calculated $K(\Delta t/t_0)$ by the eigenvalue method described in Appendix B. (The value of t_0 is

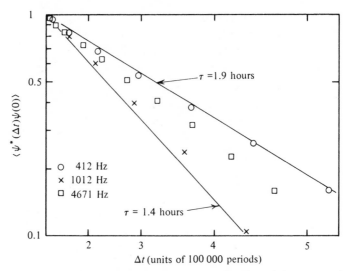

Fig. 16.8. Covariance of the wavefunction as a function of time separation. By putting the ordinate on a log scale and the abscissa on a scale that is linear in the square of Δt we obtain a straight line if (8.4.5) and (7.5.15) are valid.

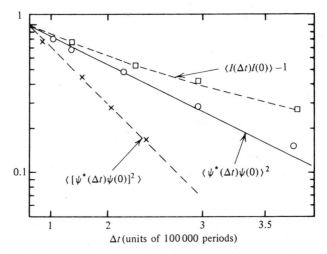

Fig. 16.9. Higher moments of the wavefunction for time separation Δt. The individual points are the experimentally measured values. In the fully saturated region the moments are expected to follow the same curve, while in the partially saturated region they follow different curves, as given in Table 16.3. The theoretical curves shown have been calculated with only one free parameter ($\hat{\Phi}$) from the internal-wave model.

Table 16.2. *Observed values of* $\dot{\Phi}$
from Fig. 16.8. The values of Φ *are*
obtained from (16.1.3). *The expected*
value of $\Phi/\dot{\Phi}$ *from the internal waves*
(7.5.16) *is* 1.4 hours.

σ (Hz)	Φ	$\dot{\Phi}$ (hours^{-1})	$\Phi/\dot{\Phi}$ (hours^{-1})
412	10.3	5.7	1.8
1012	25.3	15.8	1.5
4672	117	78	1.7

Table 16.3. *Two-point functions.*

Observed	Predicted Fully saturated	Partially saturated
$\langle I(\Delta t)I(0)\rangle - 1$	$\exp[-D(\Delta t)]$	$K(\Delta t/t_0)$
$\langle \psi^*(\Delta t)\psi(0)\rangle^2$	$\exp[-D(\Delta t)]$	$\exp[-D(\Delta t)]$
$\langle [\psi^*(\Delta t)]^2[\psi(0)]^2\rangle$	$\exp[-D(\Delta t)]$	$\exp[-2D(\Delta t)]$

0.16 hours.) The expectations from the partially saturated regime are fully borne out, verifying the phasor behavior in time illustrated in Fig. 8.10.

16.3. Pulse measurements

The short-pulse measurements have been used to measure pulse travel time distribution, including the *pulse time extent*; and the lagged intensity covariance, including the spread and wander of a pulse (§ 8.7).

Pulse extent

The distribution of arrival energy averaged over many pulses, $\langle I(\tau)\rangle$, is shown in Fig. 16.10. Our prediction for the shape of this

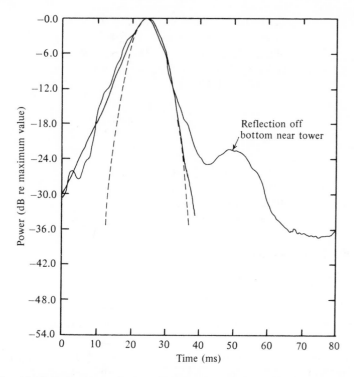

Fig. 16.10. Pulse extent in the AFAR experiment. The theoretical expectation comes from (8.7.3) which includes two effects; the exponential factor
with bandwidth σ/Φ and the Q factor with bandwidth η_0. The dashed
theoretical curve assumes η_0 is very large so that only the symmetric
exponential factor enters. The solid theoretical curve has the Q factor from
the internal-wave model included.

experimental distribution must come from (8.7.3), where $Q(\Delta\sigma/\eta_0)$
is calculated according to Appendix B. The bandwidth η_0 from
(8.6.10) is 40 Hz compared with the bandwidth in the exponential
factor of $\sigma/\Phi = 40$ Hz, thus the Q factor results in a wider pulse
extension. The predictions of the exponential factor alone and the
addition of the Q factor are indicated in Fig. 16.10. The asymmetric
shift toward early arrivals is a result of the action of the deterministic sound channel; thus this striking agreement between
experiment and theory depends crucially on a proper model for the
sound channel, provided in this case by the AFAR group's sixth-
order polynomial fit.

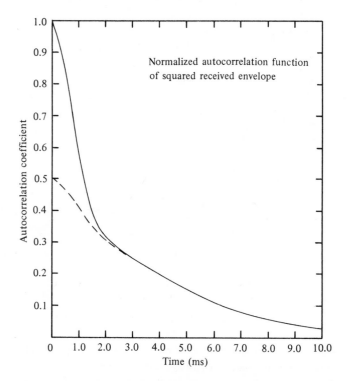

Fig. 16.11. Autocorrelation coefficient of received intensity as a function of pulse time. By subtracting the coefficient of the transmitted pulse, we have obtained the intensity correlation function for a δ-function input as the dashed curve.

Lagged covariance; spread and wander

In the absence of fluctuations the experimental lagged covariance, $F(\tau')$, from (8.7.4), would be identical to the autocorrelation of the transmitted signal. Due to fluctuations, $F(\tau')$ develops a tail (Fig. 16.11) which is proportional to the Fourier transform of $|Q(\Delta\sigma/\eta_0)|^2$.

A comparison of the Fourier transform of the experimentally observed tail and the calculation of $|Q(\Delta\sigma/\eta_0)|^2$ from the eigenvalue method described in Appendix B shows very good agreement (Fig. 16.12). Note that the pulse extent observed ($\sim$5 ms) is very similar to the spread deduced from the tail of $F(\tau')$; the differentiation between the partially and fully saturated region is much

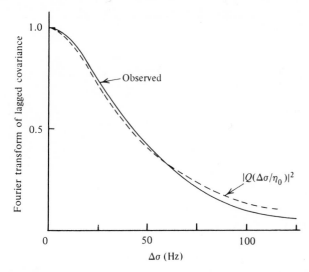

Fig. 16.12. A comparison between the Fourier transforms of the lagged covariance (dashed curve in Fig. 16.11) and the prediction $|Q|^2$ from the internal-wave model.

more sensitively made in the CW measurements (Fig. 16.9) than in the pulsed measurements. In all cases the calculations of Q based on the known sound channel, and the value of η_0 based on the calculated values of Λ and Φ, are verified by experiment.

EPILOG

The comparison between our theories and several experiments, delineated in the last three chapters, has been a first step toward relating ocean sound propagation to the physical ocean structure now evolving from oceanographic theory and experiment. The three experiments, located at Cobb Seamount, the Azores Fixed Acoustic Range, and Eleuthera–Bermuda, span the sound-transmission regimes from almost unsaturated to fully saturated (Fig. V.1). The described comparisons to theory have been confined to signal *time dependence*: measurements of coherence times and associated frequency spectra (all experiments); frequency spectra in the multipath environment (Eleuthera–Bermuda); coherent bandwidth and pulse propagation (Azores). Except for the intensity fluctuations in the Cobb Seamount experiment, the large number of experimental results have all been in agreement with predictions obtained from our theories under the assumption of internal-wave dominance, at least to the accuracy expected from taking a universal spectrum of internal waves for all the world's oceans. However, definitive experiments on signal *spatial dependence* (i.e. coherence lengths and associated spatial spectra), as well as *mixed signal dependence* on space and time together, have not yet been done. Future experiments, particularly those concerned with spatial dependence, may validate internal-wave dominance, or may lead to a useful probe of other ocean processes that prove important.

APPENDICES

CALCULATION OF $K(\alpha)$

For most cases of interest in the ocean we can write

$$K(\alpha) = |Q(\alpha)|^2 \qquad (A.1)$$

where $Q(\alpha)$ is discussed in Appendix B.

The precise definition of $K(\alpha)$ is given in terms of an ordinary differential equation for a two-by-two matrix $M_{ij}(x, \alpha)$:

$$\left[(\partial_{xx} + U'') \begin{pmatrix} 0 & 1 \\ 1 & 0 \end{pmatrix} + i\alpha F(x) \begin{pmatrix} 1 & 0 \\ 0 & C(x) \end{pmatrix} \right] M(x, \alpha) = 0 \qquad (A.2)$$

where the boundary conditions are $M(0, \alpha) = 0$ and $\partial_x M(0, \alpha) = \delta_{ij}$, and $F(x)$ is defined in Appendix B. The quantity $K(\alpha)$ is det $[M(\alpha, R)]$. The function $C(x)$ is defined in terms of the behavior of the structure function $d(\zeta, t)$ at small vertical separations and times (Chapter 7). It requires a second-order coupling between ζ and t. If

$$\partial_{\zeta\zeta} d(0, t) = \partial_{\zeta\zeta} d(0, 0)[1 - \tfrac{1}{2}(t/\tau')^2], \qquad (A.3)$$

then

$$C(x) = (\Phi/\dot{\Phi}\tau')^2. \qquad (A.4)$$

If $C(x)$ is a constant equal to unity then (A.1) holds. If $C(x)$ can be considered a constant, either because it is, or because $F(x)$ is a δ-function, then

$$K(\alpha) = |Q(C^{\frac{1}{2}}\alpha)|^2. \qquad (A.5)$$

For the GM spectrum $\tau' = \Phi/\dot{\Phi}$ for a flat ray, and for a steep ray we need only consider τ' near the apex where again it is $\Phi/\dot{\Phi}$ because the contributions to $\Phi/\dot{\Phi}$ come only from the apex region. Thus in both cases (A.1) holds.

CALCULATION OF $Q(\alpha)$

The micropath bandwidth function $Q(\alpha)$ is computed by solving an ordinary differential equation

$$-\partial_{xx}S(x, \alpha) - U_0''S(x, \alpha) = i\alpha F(x)S(x, \alpha) \tag{B.1}$$

with the boundary conditions $S(0, \alpha) = 0$, $\partial_x S(0, \alpha) = 1$, where

$$F(x) = (\pi^2/6)q_0\langle\mu^2\rangle L_P(\theta, z)(L_V^2\Lambda\Phi^2)^{-1}. \tag{B.2}$$

Then $Q(\alpha)$ is given by

$$Q(\alpha) = [S(R, 0)/S(R, \alpha)]^{\frac{1}{2}}. \tag{B.3}$$

An equivalent method which avoids solving (B.1) for every possible value of α, and also avoids the two-coupled-equation character of (A.1) resulting from the fact that S is a complex function, is to utilize eigenvalues λ_n defined such that

$$-\partial_{xx}S_n(x) - U_0''S_n(x) = \lambda_n F(x)S_n(x) \tag{B.4}$$

with the boundary conditions $S_n(0) = S_n(R) = 0$.
Then $Q(\alpha)$ is given by

$$Q(\alpha) = \prod_n (1 - i\alpha/\lambda_n)^{-\frac{1}{2}}. \tag{B.5}$$

Typically, evaluation of $Q(\alpha)$ will proceed by calculating the first few eigenvalues, λ_n, under the assumption that $F(x)$ is either a constant (for short range or a flat ray), or a δ-function near each turning point of the unperturbed ray for long-range, steep rays. When $F(x)$ is a constant it can be approximated as

$$F(x) \approx \pi^2/R^2. \tag{B.6}$$

No sound channel

In this case $U_0'' = 0$ and the problem corresponds to simple transverse waves on a string for a constant $F(x)$. Then

$$\lambda_n = n^2 \tag{B.7}$$

$$Q(\alpha) = (i\alpha\pi^2)^{\frac{1}{4}}[\sin (i\alpha\pi^2)^{\frac{1}{2}}]^{-\frac{1}{2}} \tag{B.8}$$

or

$$Q(\alpha) = \prod_n (1 - i\alpha/n^2)^{-\frac{1}{2}}. \tag{B.9}$$

Sound channel

A normal sound channel has U_0'' positive, which may pull the first one or two eigenvalues down to negative values. The sign of the eigenvalues is important only in pulse transmission experiments, because the phase of $Q(\alpha)$ determines the retardation or advance of a frequency component of a transmission. If all eigenvalues are positive, then the perturbed pulse always lags behind the time that an unperturbed pulse would arrive. However, with a sound channel that results in negative eigenvalues, significant energy will arrive in *advance* of an unperturbed arrival.

Long-range steep rays

In this case $Q(\alpha)$ reproduces exactly the form of (B.8) or (B.9) except that for some sound channels (e.g. the canonical sound channel) the quantity α should be replaced by $-\alpha$. The correct sign for α is determined by the sign of δ in (4.1.12).

Single, upper, turning point

Replace $F(x)$ by a δ-function at the turning point.

$$Q(\alpha) = (1 - i\alpha\pi^2/6)^{-\frac{1}{2}}. \tag{B.10}$$

Useful formula

$$|\sin (i\alpha\pi)^{\frac{1}{2}}|^2 = \sinh^2 (\beta) + \sin^2 (\beta)$$
$$\beta = (\tfrac{1}{2}\alpha\pi^2)^{\frac{1}{2}}. \tag{B.11}$$

APPENDIX C

CALCULATION OF γ

Assuming $L_V \ll L_H$ as in the ocean the constant γ in (12.2.3) is

$$\gamma = -\int_0^R dx \int_{-\infty}^{\infty} dk_V \, \tilde{d}(k_V, x) Q(k_V, x) \qquad (C.1)$$

where

$$\tilde{d}(k_V, x) = \frac{1}{2\pi} \int_{-\infty}^{\infty} e^{-ik_V z} d(z, x) \, dz \qquad (C.2)$$

$$Q(k_V, x) = \left[1 - \cos\left(\frac{k_V^2}{q} g(x, x) \right) \right] \exp\left[-\int_0^R d\left(\frac{k_V}{q} g(x, x'), x' \right) dx' \right] \qquad (C.3)$$

and g is defined as the solution to the differential equation

$$\partial_x^2 g(x, x') + U_0''(x) g(x, x') = \delta(x - x') \qquad (C.4)$$

with the boundary conditions $g(0, x') = g(R, x') = 0$ where $U_0''(x)$ is $\partial_z^2 U_0(z)$ evaluated along the unperturbed ray. The derivation of (C.1) to (C.4) is a straightforward generalization of the calculation (Dashen, 1977) in a homogeneous isotropic medium.

BIBLIOGRAPHY

The numbers in brackets at the end of each reference are the page numbers on which reference is made to that publication.

Adams, S. L. (1976). Multipath scintillation spectra. *J. Acoust. Soc. Am.* **60**, 1218–19. [158]

Baer, R. N., and Jacobsen, M. J. (1975). SOFAR transmission fluctuations produced by a Rossby wave. *J. Acoust. Soc. Am.* **57**, 569–76. [39]

Barabanenkov, Yu. N., Kravtsov, Yu. A., Rytov, S. M., and Tatarskii, V. I. (1970). The state of the theory of the scattering of waves in a randomly inhomogeneous medium. *Usp. Fiz. Nauk.* **102**, 1–42, in Russian. [226]

Beckmann, P. (1967). *Probability in communications engineering.* New York: Harcourt, Brace, and World. [134]

Beran, M. J. (1970). Propagation of a finite beam in a random medium. *J. Opt. Soc. Am.* **60**, 518–21. [99]

Beran, M. J. (1975). Coherence equations governing propagation through random media. *Radio Science* **10**, 15–21. [86, 226]

Beran, M. J., McCoy, J., and Adams, B. (1975). U.S. Naval Research Laboratory Report 7809. [85, 214]

Bond, J. W., Watson, K. M., and Welch, J. A. (1965). *Atomic theory of gas dynamics.* Reading, Massachusetts: Addison-Wesley. [220]

Brekhovskikh, L. M. (1960). *Waves in layered media.* New York: Academic Press. [xix]

Brekhovskikh, L. M., ed. (1975). *Acoustics of the ocean.* U.S. Joint Publication Research Service. [xix, 85]

Brekhovskikh, L. M., Ivanov-Frantskevich, G. N., Koshlyakov, M. N., Fedorov, K. N., Fomin, L. M., and Yampolsky, A. P. (1971). Some results of a hydrophysical experiment on a test range established in the tropical Atlantic Ocean. *Izvestia* **7**, 332. [34]

Bryan, K., Manabe, S., and Pacanowski, R. C. (1975). A global ocean–atmosphere climate model. Part II. The oceanic circulation, *J. Phys. Oceanogr.* **5**, 30–46. [23]

Cairns, J., and Williams, G. (1976). Internal wave measurements from a midwater float, II. *J. Geophys. Res.* **81**, 1943–50. [54]

Chandrasekhar, S. (1950). *Radiative transfer.* London: Oxford University Press. [220]

Chernov, L. A. (1960). *Wave propagation in a random medium* (translated by R. A. Silverman). New York: McGraw-Hill. [xix, 85]

Cohen, M. H., Gundermann, E. J., Hardebeck, H. E., and Sharp, L. E. (1967). Interplanetary scintillations: II. Experiment. *Astrophysical J.* **147**, 449–66. [85]

Cox, H. (1977). The approximate ray-angle diagram. *J. Acoust. Soc. Am.* **61**, 353–9. [73]

Crease, J. (1962). Velocity measurements in the deep water of the western North Atlantic. *J. Geophys. Res.* **67**, 3173–76. [34]

Darwin, C. (1844). Letter to J. M. Herbert, 1844 or 1845, in *The life and letters of Charles Darwin*, F. Darwin, ed., Vol. 1, p. 334. London, 1887: Murray. [vi]

Dashen, R. (1979). Path integrals for waves in random media. Stanford Research Institute Technical Report JSR 76-1, *Journal of Mathematical Physics*, in press. [148, 208–14]

da Vinci, Leonardo (1483). Manuscript B, Institut de France, Folio 6 recto. [Frontispiece]

Desaubies, Y. J. F. (1976). Acoustic-phase fluctuations induced by internal waves in the ocean. *J. Acoust. Soc. Am.* **60**, 795–800. [254]

de Wolf, D. A. (1975). Propagation regimes for turbulent atmospheres. *Radio Science* **10**, 53–7. [92]

Dyer, I. (1970). Statistics of sound propagation in the ocean. *J. Acoust. Soc. Am.* **48**, 337–45. [122]

Dyson, F., Munk, W. H., and Zetler, B. (1976). Interpretation of multipath scintillations Eleuthera to Bermuda in terms of internal waves and tides. *J. Acoust. Soc. Am.* **59**, 1121–33. [154–7, 239–43, 249–50]

Eckart, C. (1960). *Hydrodynamics of the oceans and atmospheres*. New York: Pergamon Press. [49]

Ellinthorpe, A. W., and Nuttall, A. H., Proceedings of the IEEE Annual Communications Convention, Boulder, Colorado, June 7–9, 1965. [144]

Ellinthorpe, A. W. *et al.* (1977). Naval Underwater Systems Center Technical Memoranda associated with the Joint Oceanographic/Acoustic Experiment (1975–77), New London, Connecticut, and private communication. [256, 260–1]

Ewart, T. E. (1976). Acoustic fluctuations in the open ocean – a measurement using a fixed refracted path. *J. Acoust. Soc. Am.* **60**, 46–59. [252–3]

Ewing, W. M., and Worzel, J. L. (1948). Long range sound transmission. *Geol. Soc. Am. Mem.* **27**, part III, 1–35. [65]

Ewing, W. M., Jardetsky, L. L., and Press, F. (1957). *Elastic waves in layered media*. New York: McGraw-Hill. [85]

Feynman, R., and Hibbs, A. (1965). *Quantum mechanics and path integrals*. New York: McGraw-Hill. [207]

Fisher, F., and Simmons, V. (1977). Sound absorption in sea water. *J. Acoust. Soc. Am.* **62**, 558–64. [232]

Flatté, S. M. (1976). Angle-depth diagram for use in underwater acoustics. *J. Acoust. Soc. Am.* **60**, 1020–3. [70–3]

Flatté, S. M., and Tappert, F. D. (1975). Calculation of the effect of internal waves on oceanic sound transmission. *J. Acoust. Soc. Am.* **58**, 1151–59. [185]

Foldy, L. L. (1945). The multiple scattering of waves I. General theory of isotropic scattering by randomly distributed scatterers. *Phys. Rev.* **67**, 107–19. [222]

Foster, T. D. (1975). Heat exchange in the upper Arctic Ocean. *AIDJEX Bulletin* **28**, 151–66. (Published by the University of Washington, Seattle.) [31]

Garrett, C., and Munk, W. H. (1972). Space-time scales of internal waves. *Geophys. Fl. Dyn.* **3**, 225–64 (referred to in the text as GM72). [49–58]

Garrett, C., and Munk, W. H. (1975). Space-time scales of internal waves: a progress report. *J. Geophys. Res.* **80**, 291–7 (referred to in the text as GM75). [54–60, 228]

Gill, A. E. (1971). The equatorial current in a homogeneous ocean. *Deep-Sea Res.* **18**, 421–31. [28]

Gill, A. E. (1973). Circulation and bottom water production in the Weddell Sea. *Deep-Sea Res.* **20**, 111–40. [11]

Gill, A. E., and Bryan, K. (1971). Effects of geometry on the circulation of a three-dimensional southern-hemisphere ocean model. *Deep-Sea Res.* **18**, 685–721. [28]

Grant, H. L., Moilliet A., and Vogel, W. M. (1968). Some observations of the occurrence of turbulence in and above the thermocline. *J. Fluid Mech.* **34**, 443–8. [15]

Gregg, M. C. (1975). Microstructure and intrusions in the California Current, *J. Phys. Oceanogr.* **5**, 253–78. [15–19]

Gregg, M. C., and Cox, C. S. (1972). The vertical microstructure of temperature and salinity. *Deep-Sea Res.* **19**, 355–76. [15]

Gregg, M. C., Cox, C. S., and Hacker, P. W. (1973). Vertical microstructure measurements in the central North Pacific. *J. Phys. Oceanogr.* **3**, 458–69. [15, 17]

Hardin, R. H., and Tappert, F. D. (1973). *SIAM Rev.* (*Chronicles*) **15**, 423. [77, 78]

Hasselmann, K. (1962). On the non-linear energy transfer in a gravity-wave spectrum Part 1. *J. Fluid Mech.* **12**, 481–500. [222]

Hasselmann, K. (1963*a*). On the non-linear energy transfer in a gravity-wave spectrum Part 2. *J. Fluid Mech.* **15**, 273–81. [222]

Hasselmann, K. (1963*b*). On the non-linear energy transfer in a gravity-wave spectrum Part 3. *J. Fluid Mech.* **15**, 385–98. [222]

Horibe, Y. ed. (1970). Preliminary report of the Hakuhō Maru Cruise KH-68-4 (Southern Cross Cruise) November 14, 1968 – March 3, 1969, Central and South Pacific. Ocean Research Institute, University of Tokyo, 170 pp. [11]

Horibe, Y. ed. (1971). Preliminary report of the Hakuhō Maru Cruise KH-70-2 (Great Bear Expedition) April 14 – June 18, 1970, North Pacific. Ocean Research Institute, University of Tokyo, 90 pp. [10]

Iselin, C. O'D. (1939). The influence of vertical and lateral turbulence on the characteristics of the waters at mid-depths. *Trans. Am. Geophys. Un.* **20**, 414–17. [9]

Kamenkovich, V. M. (1962). On the theory of the antarctic circumpolar current. *Proc. Inst. Oceanol.* **56**, 241–93. (Translated by W. D. McKee and R. Radok, 1966.) [28]

Kibblewhite, A. C., and Bedford, N. R. (1975). Regional dependence of low frequency attenuation in the ocean. *J. Acoust. Soc. Am.* **58**, S85. [232]

Kibblewhite, A. C., and Denham, D. G. (1971). Low-frequency acoustic attenuation in the South Pacific ocean. *J. Acoust. Soc. Am.* **49**, 810–15. [232]

Lau, C. W., and Watson, K. M. (1970). Radiation transport along curved ray paths. *J. Math. Phys.* **11**, 3125–37. [223]

Leontovich, M., and Fok, V. (1946). Solution of the problem of propagation of electromagnetic waves along the earth's surface by the parabolic equation method. *Zh. Eksp. Teor. Fiz.* **16**, 557–73. [77]

Mellen, R. H., and Browning, D. G. (1975). Low frequency attenuation in the Pacific Ocean. *J. Acoust. Soc. Am.* **57**, S65. [232]

Mellen, R. H., Browning, D. G., and Ross, J. M. (1974). Attenuation in randomly inhomogeneous sound channels. *J. Acoust. Soc. Am.* **56**, 80–2. [232]

Mercier, R. P. (1962). Diffraction by a screen causing large random phase fluctuations. *Proc. Cambridge Phil. Soc.* **58**, 382–400. [189]

MIMI references [p. 239].

Steinberg, J. C., and Birdsall, T. G. (1966). Underwater sound propagation in the straits of Florida. *J. Acoust. Soc. Am.* **39**, 301–15.

Steinberg, J. C., Clark, J. G., DeFerrari, H. A., Kronengold, M., and Yacoub, K. (1972). Fixed-system studies of underwater-acoustic propagation. *J. Acoust. Soc. Am.* **52**, 1521–35.

Clark, J. G., and Yarnell, J R. (1973). Propagation in time-dependent ideal channels. Application to straits of Florida studies. *J. Acoust. Soc. Am.* **53**, 556–60.

Clark, J. G., Weinberg, N. L., and Jacobson, M. J. (1973). Refracted, bottom-reflected ray propagation in a channel with time-dependent linear statification. *J. Acoust. Soc. Am.* **53**, 802–18.

Clark, J. G., and Kronengold, M. (1974). Long-period fluctuations of CW signals in deep and shallow water. *J. Acoust. Soc. Am.* **56**, 1071–83.

Munk, W. H. (1950). On the wind-driven ocean circulation. *J. Met.* **7**, 79–93. [27]

Munk, W. H. (1966). Abyssal recipes. *Deep-Sea Res.* **13**, 707–30. [7]

Munk, W. H. (1974). Sound channel in an exponentially stratified ocean, with application to SOFAR, *J. Acoust. Soc. Am.* **55**, 220–6. [5, 32, 65, 68, 234]

Munk, W. H., and Zachariasen, F. (1976). Sound propagation through a fluctuating stratified ocean–theory and observation. *J. Acoust. Soc. Am.* **59**, 818–38 (referred to in the text as MZ76). [68, 104, 110, 113, 165–86, 239–55]

Munk, W. H., and Zachariasen, F. (1977). Scattering of sound by internal waves: the role of particle velocities; to be submitted to J. Acoust. Soc. Am. [46]

Niiler, P. P. (1975). Deepening of the wind-mixed layer. *J. Mar. Res.* **33**, 405–22. [30]

Norton, K. A., Vogler, L. E., Mansfield, W. V., and Short, P. J. (1955). The probability distribution of the amplitude of a constant vector plus a Rayleigh-distributed vector. *Proc. IRE*, **43**, 1354–61. [134]

Officer, C. B. (1958). *Introduction to the theory of sound transmission.* New York: McGraw-Hill. [xviii, 63]

Pedersen, M. A. (1968). Ray theory applied to a wide class of velocity functions. *J. Acoust. Soc. Am.* **43**, 619–34. [68]

Phillips, O. M. (1966). *The dynamics of the upper ocean.* Cambridge University Press. [xviii, 49]

Pollard, R. T., Rhines, P. B. and Thompson, R. O. R. Y. (1973). The deepening of the wind-mixed layer. *Geophys. Fl. Dyn.* **3**, 381–404. [30]

POLYMODE Organizing Committee. Recommendations of November 1975 to the Joint POLYMODE Organizing Committee. Appendix II, Variability of Ocean Currents and the General Ocean Circulation: Review and Current Status of the Problem. [40, 41]

Prokhorov, A. M., Bunkin, F. V., Gochelashvily, K. S., and Shishov, V. I. (1975). Laser irradiance propagation in turbulent media. *Proc. IEEE* **63**, 790–811. [226]

Proni, J. R., and Apel, J. R. (1975). On the use of high-frequency acoustics for the study of internal waves and microstructure. *J. Geophys. Res.* **80**, 1147–51. [223]

Rattray, M., and Welander, P. (1975). A quasi-linear model of the combined wind-driven and thermohaline circulations in a rectangular β-plane ocean. *J. Phys. Oceanogr.* **5**, 585–602. [23]

Reid, J. L. (1961). On the temperature, salinity and density differences between the Atlantic and Pacific Oceans in the upper kilometer. *Deep-Sea Res.* **7**, 265–75. [14]

Reid, J. L., and Lynn, R. J. (1971). On the influence of the Norwegian–Greenland and Weddell Seas upon the bottom waters of the Indian and Pacific Oceans. *Deep-Sea Res.* **18**, 1063–88. [12, 24]

Rhines, P. B. (1975). Waves and turbulence on a beta-plane. *J. Fluid Mech.* **69**, 417–43. [35]

Rhines, P. B. (1976). The dynamics of unsteady currents. *The Sea, VI.* Wiley-Interscience, New York. [36, 39, 40, 42]

Roberts, J. (1973). University of Alaska IMS Report No. R73–4 (unpublished). [54]

Robinson, A. R. (1975). The variability of ocean currents. *Revs. Geophys. and Space Phys.* **13**, 598–602. [32]

Roden, G. I. (1975). On North Pacific temperature, salinity, sound velocity and density fronts and their relation to the wind and energy flux fields. *J. Phys. Oceanogr.* **5**, 557–71. [28–9]

Rytov, S. (1937). Wave and geometrical optics. *Comptes Rendus (Doklady) de l'Acad. des Sciences, USSR*, **18**, 263 (1938). [84, 85]

Salpeter, E. E. (1967). Interplanetary scintillations: I. Theory. *Astrophys. J.* **147**, 433–48. [85, 189]

Schulken, M., and Marsh, H. M. (1962). Sound absorption in seawater. *J. Acoust. Soc. Am.* **34**, 864–5. [232]

Sheely, M. J., and Halley, R. (1957). Measurement of the attenuation of low frequency underwater sound. *J. Acoust. Soc. Am.* **29**, 464–9. [232]

Snider, R. F. (1960). Quantum-mechanical modified Boltzmann equation for degenerate internal states. *J. Chem. Phys.* **32**, 1051–60. [222]

Stommel, H. (1958). The abyssal circulation. *Deep-Sea Res.* **5**, 80–2. [24, 25]

Stommel, H., Arons, A. B., and Faller, A. J. (1958). Some examples of stationary planetary flow patterns in bounded basins. *Tellus* **10**, 179–87. [24]

Stott, P. E. (1968). Transport equation for multiple scattering of electromagnetic waves by turbulent plasma. *J. Phys.* **A1**, 675–89. [222]

Sverdrup, H. U., Johnson, M. W., and Fleming, R. H. (1970). *The oceans, their physics, chemistry, and general biology.* New York: Prentice-Hall. [26]

Tappert, F. D. (1974). Parabolic equation method in underwater acoustics. *J. Acoust. Soc. Am.* **55**, S34. [77]

Tappert, F. D., and Hardin, R. H. (1973). In A synopsis of the AESD Workshop on Acoustic Modeling by Non-Ray Techniques, May, Washington, D.C., *AESD TN-73-05*, Office of Naval Research, Arlington. [77]

Tappert, F. D., and Hardin, R. H. (1974). In *Proceedings of the Eighth International Congress on Acoustics* Vol. II, p. 452. London: Goldcrest. [77]

Tatarskii, V. I. (1971). *The effects of the turbulent atmosphere on wave propagation.* Israel Program for Scientific Translation, Jerusalem. (Available from the National Technical Information Service, Springfield, Virginia.) [83, 85, 90–5, 99, 214]

Thorp, W. H. (1965). Deep-ocean sound attenuation in the sub- and low-kilocycle-per-second region. *J. Acoust. Soc. Am.* **38**, 648–54. [232]

Tolstoy, I., and Clay, C. S. (1966). *Ocean acoustics.* New York: McGraw-Hill. [xviii]

Turner, J. S. (1973). *Buoyancy effects in fluids.* Cambridge University Press. [xviii, 9, 16]

Urick, R. J. (1967). *Principles of underwater sound for engineers.* New York: McGraw-Hill. [xviii]

Urick, R. J. (1975). *Principles of underwater sound.* New York: McGraw-Hill. [xviii]

Veronis, G. (1969). On theoretical models of the thermocline circulation. *Deep-Sea Res.* **16** (*Supplement*), 301–23. [23]

Watson, K. M. (1960). Quantum mechanical transport theory. I. Incoherent processes. *Phys. Rev.* **118**, 886–98. [222]

Watson, K. M. (1969). Multiple scattering of electromagnetic waves in an underdense plasma. *J. Math. Phys.* **10**, 688–702. [222]

Weinberg, H. (1975). Application of ray theory to acoustic propagation in horizontally stratified oceans. *J. Acoust. Soc. Am.* **58**, 97–109. [66]

Weston, D. E. (1967). In *Institute on underwater acoustics proceedings.* V. M. Albers, ed., New York: Plenum Press. [223]

Wigner, E. P. (1932). On the quantum correction for thermodynamic equilibrium. *Phys. Rev.* **40**, 749–59. [223]

Williams, R. E., and Battestin, H. F. (1976). Time coherence of acoustic signals transmitted over resolved paths in the deep ocean. *J. Acoust. Soc. Am.* **59**, 312–28. [250]

Woods, J. D. (1968). Wave-induced shear instability in the summer thermocline. *J. Fluid Mech.,* **32**, 791–800. [15]

Wüst, G. (1955). Stromgeschwindigkeiten im Tiefen- and Bodenwasser des Atlantischen Ozeans auf Grund dynamischer Berechnung des Meteor-Profile des Deutschen Atlantischen Expedition 1925/27. *Deep-Sea Res.* **3** (Supplement), 373–97. [23]

GLOSSARY OF TERMS

abyssal – pertaining to the deep regions in the ocean

acoustic path length – the travel time of a sound pulse along a path multiplied by some constant reference sound speed

adiabatic gradient – the gradient of any quantity (e.g. temperature, density, or sound speed) that would obtain in a fluid at constant entropy

adiabatic vertical displacement – a displacement in the vertical direction of an element of water whose entropy remains constant, such as would occur under the action of internal waves

aliased – pertaining to a frequency spectrum (of a time series) whose low-frequency components have been contaminated by inadequately sampled high-frequency components

anticylonic – pertaining to the inward spiral of fluid motion toward a low pressure center (counterclockwise in the Northern Hemisphere) caused by the earth's rotation

axial ray – the horizontal sound ray that travels directly along the sound-channel axis

baroclinic – pertaining to an ocean in which density cannot be regarded as a function of pressure only – the general case

barotropic – pertaining to an ocean in which density can be regarded as a function of pressure only – a special case

beam-former type of caustic – where a beam-forming array of receivers detects the same arrival angle regardless of the initial launch angle at the source

bilinear sound-speed profile – an approximation to a sound-channel profile obtained by taking the sound speed to be linear with depth with a positive (negative) slope above (below) the sound-channel axis

buoyancy frequency – the frequency formed from the ratio of the buoyant force to the mass for a fluid element displaced from equilibrium in the vertical direction

caustic – an envelope tangent to a sequence of rays of different launch angles at a source

cyclonic – pertaining to the outward spiral of fluid motion away from a high pressure center (clockwise in the Northern Hemisphere)

deterministic rays – the set of rays determined by the equilibrium sound-speed profile in the absence of fluctuations

eddy diffusivity – a phenomenological parameter representing the transport of some passive property (such as temperature) by ill-understood processes, such as turbulence (hence the word 'eddy')

eikonal – the acoustic path length as a function of the path end-points; another term for Hamilton's characteristic function

fadeout – an interval in a time series of amplitude when the amplitude drops below a specified threshold

fine structure – structure in ocean parameters, such as temperature and salinity, with vertical scales between 1 m and 100 m

flat ray – axial ray

Fresnel condition – the condition that the size of a Fresnel zone is much less than the spatial extent of index-of-refraction inhomogeneities; a condition that geometrical optics is valid for the instantaneous pressure

Fresnel zone – consider a point on an equilibrium ray. Construct a plane at this point transverse to the equilibrium ray. If the acoustic path lengths of the rays from the source to a point P in the plane and from P to the receiver sum to less than the acoustic path length of the equilibrium ray plus one-half wavelength, then P lies within the Fresnel zone on the plane

friction velocity – the velocity scale defined by the square root of the ratio of wind stress and water density

front – a sloping boundary between two fluid masses of markedly different densities

frontogenesis – the process of creating a front

geostrophic flow – fluid motion in which the Coriolis 'force' balances the pressure gradient

gross structure – structure in ocean parameters, such as temperature and salinity, with vertical scales larger than 100 m

gyre – the major subdivisions of the wind-driven ocean circulation (e.g., the subtropical gyre; the subpolar gyre)

haline – saline

inertial frequency – 2Ω sin (latitude) where Ω is the angular velocity of the rotation of the earth

internal waves – ocean gravity waves made possible by the density increase with depth and the restoring force of gravity; thus waves internal to the volume of the ocean

ionic relaxation processes – processes involving the dissociation–reassociation of an ionic compound (such as $MgSO_4$) whose finite relaxation time leads to absorption of a sound pressure wave *isodensity surface* – a surface of constant density, called isopycnal surfaces in most oceanography texts.

mesoscale eddy – vortex-like ocean-current structures with typical length scales of 100 to 300 km

micropaths – rays (stationary-phase points) from source to receiver generated by sound-speed fluctuations, where the equilibrium profile generates only one deterministic ray

microstructure – structure in ocean parameters, such as temperature and salinity, with vertical scales smaller than 1 m

mixed layer – the relatively uniform surface layer of the ocean (10 to 100 m deep) generally regarded as mixed by wind action

multipath – the condition of having many rays (stationary-phase points) from a single source arriving at a single receiver

phase screen – a convenient model for the effect of index-of-refraction fluctuations on wave propagation, in which the phase of the wave is regarded as changed abruptly at a vertical plane (the screen)

planetary wave – a periodic disturbance in an ocean current, involving a restoring force due to the variation of Coriolis 'force' with latitude

potential gradient – the gradient of any quantity (e.g. temperature, density or sound speed) after subtraction of the adiabatic gradient. If a fluid element is displaced and allowed to reach pressure equilibrium, the potential gradient describes the difference in properties between the displaced element and its surroundings

Richardson scale – the length at which viscous energy dissipation becomes comparable to the loss of energy due to working against buoyancy. At smaller scales buoyancy becomes negligible.

Rossby wave – planetary wave

saturated region – the region of the sound fluctuation (Λ–Φ) diagram where a large number of micropaths are generated by sound-speed fluctuations. So called because the intensity fluctuations reach a limiting (saturated) distribution with variance 5.6 dB

shallow depths – generally the upper kilometer of the ocean

shear flow – fluid flow in which a gradient of average velocity is maintained

stationary-phase interaction – a process whereby the wavevector of an internal wave is perpendicular to a sound ray, allowing coupling in the stationary-phase approximation

stratification – the arrangement of ocean water in which the density increases with depth. The word does not necessarily imply layers but applies to a continuous density gradient as well

subinertial – pertaining to ocean processes with frequencies below the inertial frequency (e.g. planetary waves) *surface-limited ray* – the ray in the sound channel that just grazes the surface at its upper turning point or apex

surface water – the surface layer of the ocean that is directly conditioned by sunshine, evaporation, and precipitation, generally regarded as the top 100 to 300 m of the ocean

thermocline – a region of large temperature gradient in the vertical direction (e.g. the lower boundary of the mixed layer)

thermohaline circulation – large-scale water motion associated with such processes as heating, cooling, evaporation, the formation of ice, etc., as distinct from wind-driven circulation

undersample – the act of recording a time series at intervals that are too large to reveal important high-frequency components; this results in aliasing

upwelling – vertical motion of deep water toward the surface

UNITS, DIMENSIONS AND GLOSSARY OF SYMBOLS

Units and Dimensions

We have expended great effort to make the equations in this book dimensionally consistent. The greatest difficulty in this regard is centered on factors of 2π due to the difference between angular and cyclical quantities, such as frequency or wavenumber. We do *not* use two different symbols for angular and cyclical frequency. We take the point of view that a given frequency (such as ω_i, the inertial frequency) can be expressed in different units like any other physical quantity. When numerically evaluating a given expression, the units of the results depend on the units of the input quantities. The only subtlety to this point of view is that one does not have to retain radians as a unit. Radians (but not cycles) can be considered dimensionless. For example, suppose we calculate Φ^2 by numerically evaluating (7.4.8). If we put q_0 in cycles km^{-1}, R and L_P in km, then the units of Φ^2 will be cycles^2. Only if we put q_0 in radians km^{-1} (or km^{-1}) may we consider Φ^2 to be a dimensionless number. Admittedly this way of thinking is a convention, but we find it has the advantage of similarity to the treatment of more conventional dimensions such as length and time.

Glossary of symbols

Only symbols that appear in more than one chapter are included in this list. Nominal numerical values are given, where appropriate, in square brackets. The equation or section in which the symbol is defined is given in the right-hand column.

A	phase curvature	(7.1.3)
A	amplitude of ψ	(8.1.5)
B	scale of the sound channel [1 km]	(1.6.3)
$C(\mathbf{x})$	sound speed	(1.1.5)
$C_P(\mathbf{x})$	potential sound speed (pressure effect removed)	(1.1.5)
C_0	reference sound speed [1500 ms^{-1}]	(4.1.1)
C_1	sound speed at the sound-channel axis	(1.1.6)

$d(c)$	phase-structure function density; a local function depending on a set of local variables $\{c\}$	(7.3.2)
d_H, d_V	coherence lengths of the sound field (horizontal and vertical)	(8.5.2–3)
$D(c)$	phase-structure function depending on a set of variables $\{c\}$	(7.3.1)
$F(d)$	spectral function which is the Fourier transform of $\rho(c)$, with $\{d\}$ including the variables conjugate to those in $\{c\}$ that have been transformed	(6.1.10)
$F_1(k)$	spectral function of sound-speed fluctuations obtained by performing a Fourier transform along a line through a three-dimensional field	(6.1.15)
G	constant, relating potential sound speed to buoyancy frequency $[2.5 \, \text{s}^2 \, \text{m}^{-1}]$	(1.1.9)
$H(j)$	mode distribution	(3.4.9)
I	sound intensity; that is, $|\psi|^2$	(8.1.7)
j	internal-wave vertical eigenmode number	(3.2.18)
j_*	characteristic mode number for the internal-wave spectrum $[j_* = 3]$	(3.4.9)
$\langle j^{-1} \rangle$	average of $1/j$	(7.4.1)
$\mathbf{k}$	wavenumber of ocean fluctuations (three components)	(6.1.9)
$\mathbf{k}_H$	horizontal wavenumber of ocean fluctuations (two components)	(6.1.10)
k_H, k_V	magnitudes of horizontal and vertical components of $\mathbf{k}$	(6.1.10)
k_y	the component of $\mathbf{k}$ parallel to a sound ray	(7.3.3)
k	same as k_H in most places	(6.1.10)
$K(\alpha)$	time correlation function	(8.4.8)
l_H, l_V	coherence lengths of the phase-structure function (horizontal and vertical)	(7.5.6–7)
L_H, L_V	correlation lengths for sound-speed fluctuations (horizontal and vertical)	(6.1.11–12)
$L_P(\theta, z)$	correlation lengths of sound-speed fluctuations along direction θ at depth z. The subscript P stands for parallel	(7.2.3)
$n(z)$	buoyancy (Brunt-Vaisala) frequency	(1.1.1)
n	same as $n(z)$ in most cases	(1.1.1)

n_0	n at the ocean surface (usually extrapolated as $5.2 \times 10^{-3} \text{ s}^{-1}$)	(1.6.3)
n_1	n at the sound-channel axis	(1.6.2)
p	spectral power-law exponent of $F_1(k)$	(6.1.11)
$p(\mathbf{x}, t)$	sound pressure field	(8.1.1)
P	pressure	(1.1.10)
q	wavenumber of sound signal (usually $q(\mathbf{x})$)	§ 4.1
$q(\mathbf{x})$	mean value of local acoustic wavenumber	(6.1.4)
q_0	reference wavenumber of sound signal	(6.1.3)
$Q(\alpha)$	frequency correlation function	(8.6.9)
r	radius of curvature at the apex of a ray	(4.1.3)
R	horizontal range; same as x in most cases	
R_D	range of a double loop	(4.1.5)
R_F	radius of the first Fresnel zone	(7.1.4)
R_S	range of a single loop (between axis crossings)	(4.1.4)
R_0	single-loop range for infinitesimal angles [20.8 km]	(4.1.6)
S	salinity	(1.1.4)
$S(\mathbf{x}, \mathbf{y})$	acoustic path length from a source at $\mathbf{x}$ to a receiver at $\mathbf{y}$ along a ray or general path	(7.1.1)
S_D	the acoustic path length of a double loop	(4.1.10)
t	time	
T	temperature	
T_c	coherence time of the sound field	(8.4.9)
u	log-amplitude, that is, $\log |\psi|$	(8.1.6)
$U(\mathbf{x})$	sound-speed deviation from a constant	(6.1.1)
$U_0(z)$	sound-speed deviation from a constant caused by the deterministic sound channel	(6.1.1)
$V(\mathbf{x})$	$2q^2 \mu(\mathbf{x}, t)$ with the time dependence implicit	(6.1.5)
$W(j, k, z)$	internal-wave eigenfunction of displacement	(3.2.16)
x	horizontal coordinate in the direction of sound transmission	
X	log-pressure, that is $\log \psi$	(10.1.11)
X	Re ψ	(9.2.1)
y	horizontal coordinate perpendicular to the direction of sound transmission	
Y	Im ψ	(9.2.1)

z	vertical coordinate (positive upward) sometimes loosely called depth	
z_1	vertical position of the sound-channel axis	(1.6.2)
$\bar{z}$	average z of two points; used in defining local quantities	(6.1.8)
γ	Doppler shift (conjugate to Δt)	§ 8.1
γ	coefficient for corrections to Gaussian statistics	(12.2.3)
γ_A	adiabatic gradient of sound speed	(1.1.10)
δ	a geometrical factor dependent on the deterministic sound channel	(4.1.11)
Δt	time separation between two points	(8.1.12)
Δy	y separation between two points	(8.1.12)
Δz	vertical separation between two points	(8.1.12)
$\Delta \sigma$	frequency separation between two points	(8.1.12)
ε	strength of the deviation of the deterministic sound channel from a constant $[5.7 \times 10^{-3}]$	(1.6.6)
ζ	vertical displacement of an isodensity surface	(3.4.1)
η_W	coherent bandwidth of the sound field	(8.6.12)
θ	angle with respect to the horizontal (positive upward)	§ 4.1
θ_H, θ_V	horizontal and vertical arrival angle (conjugate to Δy and Δz)	§ 8.1
$\overline{\theta^2}$	variance of angle across a turning point	(7.4.16)
ι	log-intensity; that is, $\log |\psi|^2$	(8.1.7)
λ	general symbol for wavelength (usually acoustic wavelength so that $\lambda = 2\pi/q$)	(6.1.2)
Λ	diffraction parameter of a ray	(7.2.5)
$\mu(\mathbf{x})$	fractional sound-speed deviation from the deterministic sound channel (sometimes $\delta C/C$)	(6.1.1)
ν	same as $\dot{\Phi}$	(9.1.6)
$\xi_1(x), \xi_2(x)$	separations between two nearby rays	(7.1.8)
ρ	density of seawater	§ 1.1
$\rho(c)$	correlation function of sound-speed fluctuation where $\{c\}$ is a set of variables	(6.1.7)
σ	acoustic frequency	(6.1.2)
τ	pulse time (conjugate to $\Delta \sigma$)	§ 8.1

τ	$\Phi/\dot{\Phi}$	
ϕ	phase of ψ	(8.1.5)
Φ	strength parameter of a ray	(7.2.1)
$\dot{\Phi}$	*rms* rate-of-change of the strength parameter Φ	(7.5.15)
$\psi(\mathbf{x}, t)$	complex demodulated sound pressure field	(8.1.4)
ω	internal-wave frequency	(3.2.12)
ω_i	inertial frequency	(1.1.1)
$\partial_x, \partial_{xy} \ldots$	partial derivatives $\partial/\partial x$, $\partial^2/\partial x \partial y \ldots$	
$\langle \, \rangle$	average over a statistical ensemble of fluctuations	

INDEX

absorption
 definition of, 223
 relaxation processes in, 232
absorption coefficient, see scattering of
 rays
abyssal circulation, 23
acoustic path length
 in definition of Fresnel zone, 101
 in definition of phase curvature, 102
 of a deterministic double loop, 68
 in distinguishing regions in Λ–Φ
 space, 127
 relation to eikonal, 83
 relation to travel time, 83
acoustic wavefunction, see
 wavefunction
amplitude, complex, see wavefunction,
 reduced
angle–depth diagrams, 69–73
 source lines in, 73
angular spectrum, see scattering of
 rays
Antarctic Circumpolar Current, 11
apex, definition of, 66
apex approximation, 113
Argand diagram, see phasor
aspect ratio
 definition of, 4
 effect on signal statistics, 214
attenuation, see absorption
autocorrelation, see second-order
 correlation

baroclinic flow, 22
barotropic flow, 22
barotropic waves, see planetary waves
Boric acid sound absorption, 232
Born approximation, 194
bottom-trapped waves, 11

boundaries in Λ–Φ space, 133, 194
Boussinesq approximation, 47
Brunt–Väisälä frequency, see
 buoyancy frequency
buoyancy field, 14
buoyancy frequency
 definition of, 3
 exponential, 4, 31, 65

canonical profile
 angle–depth diagram, 69
 exponential buoyancy frequency, 4,
 31, 65
 expression for, 32
 typical midlatitude section, 14
 see also, buoyancy frequency,
 exponential
Cartesian spectra, 153, 243–8
caustic, 66, 73
 for a beam-former receiver 106,
 188
channeled ocean, 174
characteristic scales of acoustic signal
 fluctuations, 123
coherence, in frequency, 205
coherence function, see correlation
 function
coherence length, 139–40
coherence time, 138
coherent bandwidth, 142
coherent communications channel,
 143
continuity equation, 47
coordinate system, 67
correlation function
 of sound-speed fluctuations, 88
 of wavefunction, 152, 203
 see also second-order correlation

correlation lengths
L_V, L_H, 89, 110
L_P, 107, 110
Cromwell current, 28
cross-correlation, *see* second-order
correlation

diffraction parameter, Λ
definition in homogeneous, isotropic
case, 90
evaluation from internal-wave
dominance, 114–16
general definition of, 108
in phase-screen treatment, 193
in unsaturated region, 169
dispersion relation of internal waves,
50
use in changing spectral function
variables, § 7–9
Doppler shift, 123
double-diffusive processes, 17

eddies, *see* mesoscale
eddy diffusivity, vertical, 7, 23
eikonal, 83
energy flux, 220
Equatorial Undercurrent, 28
Euler equation, 47
exponential ocean, *see* canonical
profile
extinction length, 95

fadeouts, 154, 161
fast Fourier transform, 78
finestructure, 14–19
fourth moment, 209–12
see also intensity correlations
Fredholm determinant, *see* functional
determinant
Fresnel condition, 169, 174, 176
Fresnel zone, 90, 101
friction velocity, 30
front, 1, 28
subarctic, 29
frontogenesis, 29
frozen ocean assumption, 144
fully saturated, 129, 200
functional determinant, 217

Gaussian random variable, 150
Gaussian statistics
behavior of the wavefunction in the
unsaturated region, 200
first corrections to, 210, 261
geometrical optics
approximation to the pressure, 94,
167
behavior of phase and amplitude
statistics, 136, 169
in a homogeneous, isotropic
medium, 94–9
rays in the presence of arbitrary
sound-speed behavior, 82
rays in the sound channel, 65
as a region of Λ–Φ space, 130
geostrophic flow, 42
geostrophic relation, 22
geostrophic turbulence, 42–3

Helmholtz equation, 77

impulse response, 144
inertial frequency, definition of, 3
instantaneous transfer function, 144
intensity coherence time, 138
intensity correlations, 136, 203, 210
intensity fluctuations
in Azores experiment, 261, 266
in Cobb Seamount experiment,
253–5
in Eleuthera–Bermuda experiment,
244
homogeneous, isotropic case, 98
numerical experiments, 187
in partial saturation, 217
spectra, 183
statistics, 154
in unsaturated region, 169, 182
intensity in transport theory, 220
internal waves
basic equations, 46
boundary conditions, 49
dispersion relation, 50, 58
displacement of isodensity surface,
55
energy density, 52, 55
linearization, 48

internal waves—*cont.*
  phase speed, 50
  role in this book, 1
  sound-speed perturbations, 32, 59–61, 257
  spectrum of displacement, 54–7
  spectrum of sound-speed deviation, 59–61
  universal spectrum, 1
  wavefunctions, 50, 53

lateral fluxes, 9
Liouville's theorem, 73
log-amplitude fluctuations, 132, 136, 168, 253
  see also intensity fluctuations
log-normal distribution, 131, 261

mesoscale
  in the ocean environment, 1
  orders of magnitude, 40–2
  relation to scales, 4
micromultipath, *see* micropath
micropath, 129
  bandwidth, 142
  in path-integral treatment, 208
  in phase-screen treatment, 189
microstructure
  definition of, 14–19
  in the ocean environment, 1
  scattering from, 234–6
mixed layer, 1, 30
moment equations
  relation to path-integral technique, 220, 226
  relation to transport equation, 220, 223, 226
moments of the wavefunction, 151
  see also correlation function, *and* wavefunction
multipath
  deterministic, 66, 73
  distinctness of multiple rays, 129, 194

normal distribution, *see* Gaussian statistics

ocean structure
  difference from homogeneous, isotropic turbulence, 100
  intrusive features, 17
  latitudinal variation, 10
  low frequency, *see* planetary waves
  see also finestructure, internal waves, *and* microstructure
one-point function, 122, 130–5, 158

parabolic approximation, 77
parabolic wave equation, 78, 212
partial saturation
  definition of, 129
  distinguishing from full saturation, 135–138, 263
  in phase-screen treatment, 196
  time separations, 160
path integral
  introduction, 78–81
  use in calculating signal statistics, 208–9
phase curvature
  approximations to, 103
  definition of, 102–3
  in diffraction parameter, Λ, 108
  in the phase-screen treatment, 190
  in the unsaturated region, 176
phase fluctuations
  in Cobb Seamount experiment, 253
  in Eleuthera–Bermuda experiment, 244
  in the unsaturated region, 167, 182
  numerical experiments, 185
  spectra, 183
phase screen, 189
  expression for the reduced wavefunction, 191
  and path integral, 207
phase statistics, 154
phase-structure function
  evaluation from internal-wave dominance, 117–19
  general definition of, 108
  homogeneous, isotropic case, 95–8
phasor, 133, 265
planetary waves
  baroclinic, 37
  barotropic, 36

planetary waves—*cont.*
  bottom-trapped, 36
  effect on sound speed, 39
  energy flux considerations, 42
  in the ocean environment, 4
  potential gradient, 4
  potential temperature, 13
  pulse time, 123, 145
  pulse time extent, 145, 263
  pulse transmission, 144–9
    asymmetric shift toward early
      arrivals, 266

range–frequency space, 130
Rayleigh distribution, 122, 133, 261
rays, 65–9
  in path integral, 82
  in phase-screen treatment, 196
  ray-tube, 171
reduced wavefunction, *see*
    wavefunction, acoustic
regions in Λ–Φ space, 133, 194
relaxation processes in absorption, 232
Richardson scale, 15
Rossby waves, *see* planetary waves
rotation of the earth, 3
Rytov approximation, 130, 167
  in Cobb seamount experiment, 254
  intensity fluctuations, 170
  in phase-screen treatment, 195
  region of validity, 168
  see also unsaturated region

salinity minimum, 12
saturated regions, 128
  in path-integral treatment, 209
  in phase-screen treatment, 194–6
  one-point functions, 132
  see also partial saturation *and* fully
    saturated
scattering of rays
  acceptance cone, 229
  angles, 227–9
  cross-section, 223
  effective absorption coefficient, 230,
    235
scattering loss, 231
  from internal waves, 228
  from microstructure, 234
  from sound channel, 232

scintillation index, *see* intensity
    fluctuations
second moment, *see* second-order
    correlation
second-moment equation, *see* moment
    equations
second-order correlation
  in path-integral treatment, 212
  in phase-screen treatment, 199
  in the unsaturated region, 135
  under Gaussian statistics, 152
  see also correlation function,
    moment, wavefunction
separation variables (space, time and
    frequency), 122
shadow zones, 66
smooth perturbations, method of, *see*
    Rytov approximation
sound axis, depth of, 32
sound channel
  physical processes leading to, 6
  see also sound-speed profile,
    canonical profile
sound-speed minimum, *see* sound
    channel
sound-speed perturbation due to
    internal waves, *see* internal waves
sound-speed profile
  behavior with latitude and
    longitude, 13
  effect of planetary waves, 39
  potential, 6
  see also canonical profile
source lines, 73
speckle interferometry (relation to
    spread and wander), 149
spectra
  Cartesian 153, 243–8
  phase and amplitude, 155, 249
spectral functions, 88
  one dimensional, 90
split-step algorithm, 79
sporadic rays, *see* micropath
spread
  in Azores experiment, 267
  calculation of, 145
  definition of, 125
  due to deterministic multipath, 148
spread function, 147

stability frequency, *see* buoyancy
  frequency
stationarity, 124
stationary-phase approximation
  leading to rays, 82
  in the saturated region, 194
  relation to steepest descent, 203
stationary-phase points
  criteria for distinctness, 194
statistics of intensity and phase
  in Gaussian statistics, 154
  in saturated region, 132
  in unsaturated region, 131
steepest-descent integration, 203
stratification, 3, 6
strength parameter, Φ
  definition in homogeneous, isotropic
    case, 91
  evaluation from internal-wave
    dominance, 110–4
  general definition of, 106
  in phase-screen treatment, 192
  supereikonal approximation, 166
subinertial flow, *see* planetary waves,
    eddies
surface mixed layer, 1, 30
Sverdrup transport, 26

temperature–salinity relation, 9
thermohaline circulation, 22
three-point function, 123, 160
time-bandwidth product, 143
transport of ocean water, 19–29
travel-time fluctuations
  in phase-screen treatment, 196
  in unsaturated regions, *see* phase
    fluctuations
turbulence
  geostrophic, 42–3

isotropic, 15
spectrum, 90
two-point function, 122, 158

unsaturated region, 130
  in phase-screen treatment, 195
upwelling, 7, 23

vertical fluxes, 6

wander
  in Azores experiment, 267
  calculation of, 145
  definition of, 125
water masses, 10–14
  Antarctic Bottom Water, 8, 11, 12
  Antarctic Circumpolar Water, 6, 12
  Antarctic Intermediate Water, 12
  bottom water, 6
  North Atlantic Deep Water, 12
  Subarctic Intermediate Water, 12
wave equation, acoustic
  derivation, 74–8
  in unsaturated region, 165–170
wavefunction, acoustic
  correlation functions, *see* correlation
    functions
  cross-correlation, *see* second-order
    correlation
  path-integral form, 208
  phase-screen form, 192
  probability distribution, 133
  ray approximation, 82
  reduced, 121
wave parameter, 91
Western Boundary Current, 25
wind-driven circulation, 22, 24
wind stress, 26, 47